Geometry in Advanced Pure Mathematics

LTCC Advanced Mathematics Series

Series Editors: Shaun Bullett *(Queen Mary University of London, UK)*
Tom Fearn *(University College London, UK)*
Frank Smith *(University College London, UK)*

Published

Vol. 1   Advanced Techniques in Applied Mathematics
edited by Shaun Bullett, Tom Fearn & Frank Smith

Vol. 2   Fluid and Solid Mechanics
edited by Shaun Bullett, Tom Fearn & Frank Smith

Vol. 3   Algebra, Logic and Combinatorics
edited by Shaun Bullett, Tom Fearn & Frank Smith

Vol. 4   Geometry in Advanced Pure Mathematics
edited by Shaun Bullett, Tom Fearn & Frank Smith

Vol. 5   Dynamical and Complex Systems
edited by Shaun Bullett, Tom Fearn & Frank Smith

Vol. 6   Analysis and Mathematical Physics
edited by Shaun Bullett, Tom Fearn & Frank Smith

LTCC Advanced Mathematics Series — Volume 4

Geometry in Advanced Pure Mathematics

Editors

Shaun Bullett
Queen Mary University of London, UK

Tom Fearn
University College London, UK

Frank Smith
University College London, UK

NEW JERSEY · LONDON · SINGAPORE · BEIJING · SHANGHAI · HONG KONG · TAIPEI · CHENNAI · TOKYO

Published by

World Scientific Publishing Europe Ltd.
57 Shelton Street, Covent Garden, London WC2H 9HE
Head office: 5 Toh Tuck Link, Singapore 596224
USA office: 27 Warren Street, Suite 401-402, Hackensack, NJ 07601

Library of Congress Cataloging-in-Publication Data
Names: Bullett, Shaun, 1947– editor. | Fearn, T., 1949– editor. |
　　Smith, F. T. (Frank T.), 1948– editor.
Title: Geometry in advanced pure mathematics / edited by Shaun Bullett
　　(Queen Mary University of London, UK), Tom Fearn (University College London, UK),
　　Frank Smith (University College London, UK).
Description: New Jersey : World Scientific, 2017. |
　　Series: LTCC advanced mathematics series ; volume 4
Identifiers: LCCN 2016036917 | ISBN 9781786341068 (hc : alk. paper)
Subjects: LCSH: Geometry, Algebraic. | Geometry, Hyperbolic. | Geometry.
Classification: LCC QA564 .G527 2017 | DDC 516.3/5--dc23
LC record available at https://lccn.loc.gov/2016036917

British Library Cataloguing-in-Publication Data
A catalogue record for this book is available from the British Library.

Copyright © 2017 by World Scientific Publishing Europe Ltd.

All rights reserved. This book, or parts thereof, may not be reproduced in any form or by any means, electronic or mechanical, including photocopying, recording or any information storage and retrieval system now known or to be invented, without written permission from the Publisher.

For photocopying of material in this volume, please pay a copying fee through the Copyright Clearance Center, Inc., 222 Rosewood Drive, Danvers, MA 01923, USA. In this case permission to photocopy is not required from the publisher.

Desk Editors: V. Vishnu Mohan/Mary Simpson

Typeset by Stallion Press
Email: enquiries@stallionpress.com

Printed in Singapore

Preface

The London Taught Course Centre (LTCC) for PhD students in the Mathematical Sciences has the objective of introducing research students to a broad range of topics. For some students, some of these topics might be of obvious relevance to their PhD projects, but the relevance of most will be much less obvious or apparently non-existent. However, all of us involved in mathematical research have experienced that extraordinary moment when the penny drops and some tiny gem of information from outside one's immediate research field turns out to be the key to unravelling a seemingly insoluble problem, or to opening up a new vista of mathematical structure. By offering our students advanced introductions to a range of different areas of mathematics, we hope to open their eyes to new possibilities that they might not otherwise encounter.

Each volume in this series consists of chapters on a group of related themes, based on modules taught at the LTCC by their authors. These modules were already short (five two-hour lectures) and in most cases the lecture notes here are even shorter, covering perhaps three-quarters of the content of the original LTCC course. This brevity was quite deliberate on the part of the editors — we asked the authors to confine themselves to around 35 pages in each chapter, in order to allow as many topics as possible to be included in each volume, while keeping the volumes digestible. The chapters are "advanced introductions", and readers who wish to learn more are encouraged to continue elsewhere. There has been no attempt to make the coverage of topics comprehensive. That would be impossible in any case — any book or series of books which included all that a PhD student in mathematics might need to know would be so large as to be totally unreadable. Instead what we present in this series is a cross-section of some of the topics, both classical and new, that have appeared in LTCC modules in the nine years since it was founded.

The present volume covers a range of topics related to geometry. The companion volumes in this series are also recommended for further aspects of geometry and its applications: in particular the chapters on 'Differential Geometry and Mathematical Physics' and 'Noncommutative Differential Geometry' in the 'Analysis and Mathematical Physics' volume could just as easily have been included here. The main readers of the books in this series are likely to be graduate students and more experienced researchers. Whatever your mathematical background, we encourage you to dive in, and we hope that you will enjoy the experience of widening your mathematical knowledge by reading these concise introductory accounts written by experts at the forefront of current research.

Shaun Bullett, Tom Fearn, Frank Smith

Contents

Preface		v
Chapter 1.	Algebraic Geometry *Ivan Tomašić*	1
Chapter 2.	Introduction to the Modular Group and Modular Forms *W. J. Harvey*	39
Chapter 3.	Geometric Group Theory *Ian Chiswell*	85
Chapter 4.	Holomorphic Dynamics and Hyperbolic Geometry *Shaun Bullett*	125
Chapter 5.	Minimal Surfaces and the Bernstein Theorem *T. Bourni and G. Tinaglia*	167
Chapter 6.	Syzygies and Minimal Resolutions *F. E. A. Johnson*	193

Chapter 1

Algebraic Geometry

Ivan Tomašić

School of Mathematical Sciences
Queen Mary University of London
London E1 4NS, UK
i.tomasic@qmul.ac.uk

Some of the most beautiful chapters of pure mathematics of the 20th century were motivated by considerations around Weil conjectures for zeta functions of algebraic varieties over finite fields. The purpose of these notes is to introduce the reader to enough algebraic geometry to be able to understand some elementary aspects of their proof for the case of algebraic curves and appreciate their number-theoretic consequences.

1. Introduction

Intuitively speaking, algebraic geometry is the study of geometric shapes that can be locally (piecewise) described by polynomial equations.

The main advantage of restricting our attention to polynomial expressions is their versatility, given that they make sense in completely arbitrary rings and fields, including the ones which carry no intrinsic topology. This gives a 'universal' geometric intuition in areas where classical geometry and topology fail.

Consequently, methods of algebraic geometry apply in a range of mathematical disciplines, depending on the choice of rings or fields in which to solve our polynomial systems. In geometry we typically work over $\mathbb{R}$ or $\mathbb{C}$, while in number theory we choose arithmetically significant structures, such as $\mathbb{Z}$ or $\mathbb{Q}$ (in Diophantine geometry), number fields, p-adic numbers, or even structures of positive characteristic and finite fields.

We focus on the problem of counting points on varieties over finite fields and the associated Weil conjectures in Section 4. We prove the rationality conjecture in an elementary way, as well as using étale cohomology as a 'black-box', and we outline the approach to the functional equation.

1.1. *Solving polynomial equations*

Let $f(x, y)$ be a polynomial over a ring k, and let X be the 'plane curve' defined by the equation $f(x, y) = 0$. For any ring R extending k, we can consider the solution set

$$X(R) = \{(a, b) \in R \times R : f(a, b) = 0 \text{ in } R\}.$$

Example 1.1 (Diophantine geometry). Let X be a regular/smooth projective curve of genus g (Definitions 3.7 and 3.39) defined over $k = \mathbb{Q}$. Intuitively, g is the number of 'holes' in $X(\mathbb{C})$ considered as a real surface. Questions about the rational points $X(\mathbb{Q})$ immediately lead to the most difficult theorems of number theory. To informally restate [9, Theorem 30], assume $X(\mathbb{Q}) \neq \emptyset$. If $g = 0$, then $X(\mathbb{Q})$ is mostly parametrised by rational functions. If $g = 1$, then X is an elliptic curve and the Mordell–Weil theorem yields that $X(\mathbb{Q})$ is a finitely-generated group. If $g > 1$, then the Mordell conjecture/Faltings theorem states that $X(\mathbb{Q})$ is finite.

Example 1.2. Consider a plane curve X defined by $x^2 + y^2 - 1 = 0$. Over $\mathbb{R}$, this defines a circle (but note that a slightly modified 'curve' $x^2 + y^2 + 1 = 0$ has no real points). Over $\mathbb{C}$, X is again a quadratic curve, although it may be difficult to imagine (as the complex plane has real dimension 4).

Illustrating the above principle for $g = 0$, $X(\mathbb{Q})$ is infinite and given by Pythagorean triples, e.g., $(3/5, 4/5) \in X(\mathbb{Q})$.

Question 1.3. *When X is a plane curve over a finite field, what can we say about $X(\mathbb{F}_q)$? Certainly $|X(\mathbb{F}_q)| \leq q \cdot q = q^2$, but this is a crude bound. In our discussion of Weil conjectures/Riemann hypothesis for varieties over finite fields in Section 4, we will see that $|X(\mathbb{F}_q)|$ grows approximately as q.*

1.2. *Historical development*

The *naïve algebraic geometry* proposes to study algebraic *varieties* through their sets of points in an *algebraically closed field*. It was pioneered by the Italian school including Castelnuovo, Enriques, Severi. Their intuitive approach culminated in a classification of algebraic surfaces, but reached unsurmountable obstacles due to lack of rigour. The American school led by Chow, Weil and Zariski gave solid algebraic foundation to the above.

The modern theory of *schemes* and *cohomology* in algebraic geometry was forged by the French school including Serre, Grothendieck and Deligne. It works over arbitrary fields or rings, and its machinery automatically performs all the necessary bookkeeping lacking in the naïve approach.

2. Varieties and Schemes

2.1. *Affine varieties*

In this subsection, k denotes an *algebraically closed* field, and we write $A = k[x_1, \ldots, x_n]$ for the polynomial ring in n variables over k.

Definition 2.1.

- The *affine n-space* is $\mathbb{A}_k^n = \{(a_1, \ldots, a_n) : a_i \in k\}$.
- We consider an $f \in A$ as a function $f : \mathbb{A}_k^n \to k$ as follows. For $P = (a_1, \ldots, a_n) \in \mathbb{A}^n$, we let $f(P) = f(a_1, \ldots, a_n)$.

Definition 2.2. For $f \in A$, we define
$$V(f) = \{P \in \mathbb{A}^n : f(P) = 0\} \quad \text{and} \quad D(f) = \mathbb{A}^n \setminus V(f).$$
More generally, for any subset $E \subseteq A$, let
$$V(E) = \{P \in \mathbb{A}^n : f(P) = 0 \text{ for all } f \in E\} = \bigcap_{f \in E} V(f).$$

Proposition 2.3. *For $E, E', E_\lambda \subseteq A$, we have*

(1) $V(0) = \mathbb{A}^n$, $V(1) = \emptyset$;
(2) $E \subseteq E'$ *implies* $V(E) \supseteq V(E')$;
(3) $V\left(\bigcup_\lambda E_\lambda\right) = V(\sum_\lambda E_\lambda) = \bigcap_\lambda V(E_\lambda)$;
(4) $V(EE') = V(E) \cup V(E')$;
(5) $V(E) = V(\sqrt{(E)})$, *where (E) is an ideal of A generated by E and $\sqrt{\cdot}$ denotes the* radical *of an ideal,* $\sqrt{I} = \{a \in A : a^n \in I \text{ for some } n \in \mathbb{N}\}$.

This shows that sets of the form $V(E)$ for $E \subseteq A$ (called *algebraic sets*) are closed sets of a topology on $\mathbb{A}^n$, which we call the *Zariski topology*. Note that sets of the form $D(f)$ are basic open.

Example 2.4. Algebraic subsets of $\mathbb{A}^1$ are just finite sets. Thus any two open subsets intersect, and the topology is far from being Hausdorff.

Proof. As $A = k[x]$ is a principal ideal domain, every ideal $\mathfrak{a}$ in A is principal, $\mathfrak{a} = (f)$, for $f \in A$. Since k is algebraically closed, f splits in k,
$$f(x) = c(x - a_1) \cdots (x - a_n).$$
Thus $V(\mathfrak{a}) = V(f) = \{a_1, \ldots, a_n\}$. □

Definition 2.5. An *affine algebraic variety* is a closed subset of $\mathbb{A}^n$, together with the induced Zariski topology.

Definition 2.6. Let $Y \subseteq \mathbb{A}^n$ be an arbitrary set (not necessarily closed). The *ideal* of Y in A is
$$I(Y) = \{f \in A : f(P) = 0 \text{ for all } P \in Y\}.$$

Proposition 2.7. *For arbitrary $Y, Y', Y_\lambda \subseteq \mathbb{A}^n$,*

(1) $Y \subseteq Y'$ *implies* $I(Y) \supseteq I(Y')$;
(2) $I\left(\bigcup_\lambda Y_\lambda\right) = \bigcap_\lambda I(Y_\lambda)$;
(3) $V(I(Y)) = \overline{Y}$, *the Zariski closure of Y in $\mathbb{A}^n$*;
(4) *for any* $E \subseteq A$, $I(V(E)) = \sqrt{(E)}$.

Proof. For (3), note that $V(I(Y))$ is closed and contains Y. Conversely, if $V(E) \supseteq Y$, then, for every $f \in E$, $f(y) = 0$ for every $y \in Y$, so $f \in I(Y)$, thus $E \subseteq I(Y)$ and $V(E) \supseteq V(I(Y))$.

Statement (4) is commonly known as *Hilbert's Nullstellensatz*. Let us write $\mathfrak{a} = (E)$. It is clear that $\sqrt{\mathfrak{a}} \subseteq I(V(\mathfrak{a}))$. For the converse inclusion, we shall assume the *weak Nullstellensatz* (in $(n+1)$ variables, see [9, Theorem 16]): for a proper ideal J in $k[x_0, \ldots, x_n]$, we have $V(J) \neq 0$ (it is crucial here that k is algebraically closed). For any $f \in I(V(\mathfrak{a}))$, the ideal $J = (1 - x_0 f) + \mathfrak{a} k[x_0, \ldots, x_n]$ in $k[x_0, \ldots, x_n]$ has no zero in k^{n+1} so we conclude $J = (1)$, i.e., $1 \in J$. It follows (by substituting $1/f$ for x_0 and clearing denominators) that $f^N \in \mathfrak{a}$ for some N. For a complete proof see [1]. □

Corollary 2.8. *The set $D(f)$ is quasi-compact (not Hausdorff).*

Proof. If $\bigcup_i D(f_i) = D(f)$, then $V(f) = \bigcap_i V(f_i) = V(\{f_i : i \in I\})$, so $f \in \sqrt{\{f_i : i \in I\}}$, and there is a finite $I_0 \subseteq I$ with $f \in \sqrt{\{f_i : i \in I_0\}}$. □

Corollary 2.9. *There is a one-to-one inclusion-reversing correspondence*
$$Y \longmapsto I(Y)$$
$$V(\mathfrak{a}) \longmapsfrom \mathfrak{a}$$
between algebraic sets and radical ideals.

Given a point $P = (a_1, \ldots, a_n) \in \mathbb{A}^n$, the ideal $\mathfrak{m}_P = I(P)$ is maximal (because the set $\{P\}$ is minimal), and $\mathfrak{m}_P = (x_1 - a_1, \ldots, x_n - a_n)$. Weak Nullstellensatz tells us that every maximal ideal is of this form. Thus,

$$I(V(\mathfrak{a})) = \bigcap_{P \in V(\mathfrak{a})} I(P) = \bigcap_{P \in V(\mathfrak{a})} \mathfrak{m}_P = \bigcap_{\substack{\mathfrak{m} \supseteq \mathfrak{a} \\ \mathfrak{m} \text{ maximal}}} \mathfrak{m}.$$

On the other hand, in an arbitrary commutative ring,

$$\sqrt{\mathfrak{a}} = \bigcap_{\substack{\mathfrak{p} \supseteq \mathfrak{a} \\ \mathfrak{p} \text{ prime}}} \mathfrak{p}.$$

Hence, Nullstellensatz in fact claims that the two intersections coincide, i.e., that A is a *Jacobson ring*.

Definition 2.10. If Y is an affine variety, its *affine coordinate ring* is $\mathcal{O}(Y) = A/I(Y)$.

The k-algebra $\mathcal{O}(Y)$ should be thought of as the ring of polynomial functions $Y \to k$. Indeed, two polynomials $f, f' \in A$ define the same function on Y if and only if $f - f' \in I(Y)$.

Remark 2.11. If Y is an affine variety, $\mathcal{O}(Y)$ is clearly a finitely generated k-algebra. Conversely, any finitely generated *reduced* (no nilpotent elements) k-algebra is a coordinate ring of an irreducible affine variety.

Indeed, suppose B is generated by $b_1, \ldots, b_n$ as a k-algebra, and define a morphism $A = k[x_1, \ldots, x_n] \to B$ by $x_i \mapsto b_i$. Since B is reduced, the kernel is a radical ideal $\mathfrak{a}$, so $B = \mathcal{O}(V(\mathfrak{a}))$.

Remark 2.12. Let $\mathrm{Specm}(B)$ denote the set of all maximal ideals of B. Then we have one-to-one correspondences between the following sets:

(1) (points of) Y;
(2) $Y(k) := \mathrm{Hom}_k(\mathcal{O}(Y), k)$;
(3) $\mathrm{Specm}(\mathcal{O}(Y))$;
(4) maximal ideals in A containing $I(Y)$.

Indeed, let $P \in Y$, $P = (a_1, \ldots, a_n)$. We know that $I(P) \supseteq I(Y)$, so the morphism $a: \mathcal{O}(Y) = A/I(Y) \to k$, $x_i + I(Y) \mapsto a_i$ is well defined. Since the range is a field, $\mathfrak{m}_P = \ker(a)$ is maximal in $\mathcal{O}(Y)$, and its preimage in A is exactly $I(P) = \{f \in A : f(P) = 0\}$.

Definition 2.13. A topological space X is *irreducible* if it cannot be written as the union $X = X_1 \cup X_2$ of two proper closed subsets.

Proposition 2.14. *An algebraic variety Y is irreducible if and only if its ideal is prime if and only if $\mathcal{O}(Y)$ is a domain.*

Proof. Suppose Y is irreducible, and let $fg \in I(Y)$. Then
$$Y \subseteq V(fg) = V(f) \cup V(g) = (Y \cap V(f)) \cup (Y \cap V(g)),$$
both being closed subsets of Y. Since Y is irreducible, we have $Y = Y \cap V(f)$ or $Y = Y \cap V(g)$, i.e., $Y \subseteq V(f)$ or $Y \subseteq V(g)$, i.e., $f \in I(Y)$ or $g \in I(Y)$. Thus $I(Y)$ is prime.

Conversely, let $\mathfrak{p}$ be a prime ideal and suppose $V(\mathfrak{p}) = Y_1 \cup Y_2$. Then $\mathfrak{p} = I(Y_1) \cap I(Y_2) \supseteq I(Y_1)I(Y_2)$, so we have $\mathfrak{p} = I(Y_1)$ or $\mathfrak{p} = I(Y_2)$, i.e., $Y_1 = V(\mathfrak{p})$ or $Y_2 = V(\mathfrak{p})$, and we conclude that $V(\mathfrak{p})$ is irreducible. $\square$

Example 2.15.

- Space $\mathbb{A}^n = V(0)$ is irreducible as 0 is a prime ideal in the domain A.
- If $P = (a_1, \ldots, a_n) \in \mathbb{A}^n$, then $\{P\} = V(\mathfrak{m}_P)$ and $\mathfrak{m}_P = (x_1 - a_1, \ldots, x_n - a_n)$ is a maximal ideal, hence prime, so $\{P\}$ is irreducible.
- Let $f \in A = k[x, y]$ be an irreducible polynomial. Then $V(f)$ is an irreducible variety (*affine curve*); the ideal (f) is prime since A is an unique factorisation domain.
- The 'cross' $V(x_1 x_2) = V(x_1) \cup V(x_2)$ is connected but not irreducible.

Definition 2.16. A topological space X is called *noetherian*, if it has the *descending chain condition* on closed subsets: any descending sequence $Y_1 \supseteq Y_2 \supseteq \cdots$ of closed subsets eventually stabilises, i.e., there is an $r \in \mathbb{N}$ such that $Y_r = Y_{r+i}$ for all $i \in \mathbb{N}$.

Proposition 2.17. *In a noetherian topological space X, every non-empty closed subset Y can be expressed as an irredundant finite union*
$$Y = Y_1 \cup \cdots \cup Y_n$$
of irreducible closed subsets Y_i (irredundant means that $Y_i \not\subseteq Y_j$ for $i \neq j$). The sets Y_i are uniquely determined (up to reordering), and we call them the irreducible components *of Y.*

Definition 2.18. A ring R is *noetherian* if it satisfies any of the following three equivalent conditions:

(1) R has the *ascending chain condition* on ideals: every ascending chain $I_1 \subseteq I_2 \subseteq \cdots$ of ideals is stationary (eventually stabilises);
(2) every non-empty set of ideals in R has a maximal element;
(3) every ideal in R is finitely generated.

Theorem 2.19 (Hilbert's Basis Theorem). *If R is noetherian, then the polynomial ring $R[x_1, \ldots, x_n]$ is noetherian.*

Corollary 2.20. *If R is noetherian and B is a finitely generated R-algebra, then B is also noetherian.*

Remark 2.21. This means that any algebraic variety $Y \subseteq \mathbb{A}^n$ is in fact a set of solutions of a *finite* system of polynomial equations:
$$f_1(x_1, \ldots, x_n) = 0$$
$$\vdots$$
$$f_m(x_1, \ldots, x_n) = 0$$

Corollary 2.22. *Every affine algebraic variety is a noetherian topological space and can be expressed uniquely as an irredundant union of irreducible varieties.*

Proof. The algebra $\mathcal{O}(Y)$ is finitely generated over a field k, which is trivially noetherian, so $\mathcal{O}(Y)$ is a noetherian ring. A descending chain of closed subsets $Y_1 \supseteq Y_2 \supseteq \cdots$ in Y gives rise to an ascending chain of ideals $I(Y_1) \subseteq I(Y_2) \subseteq \cdots$ in $\mathcal{O}(Y)$, which must be stationary. Thus the original chain of closed subsets must be stationary too. □

Definition 2.23. The *dimension* of a topological space X is the supremum of all n such that there exists a chain
$$Z_0 \subset Z_1 \subset \cdots \subset Z_n$$
of distinct irreducible closed subsets of X.

The *dimension* of an affine variety is the dimension of its underlying topological space.

Note that not every noetherian space is finite dimensional.

Definition 2.24. In a ring R, the *height* of a prime ideal $\mathfrak{p}$ is the supremum of lengths n of chains $\mathfrak{p}_0 \subset \mathfrak{p}_1 \subset \cdots \subset \mathfrak{p}_n = \mathfrak{p}$ of distinct prime ideals.

The *Krull dimension* of R is the supremum of the heights of all the prime ideals.

Fact 1. Let B be a finitely generated k-algebra which is a domain. Then
(1) $\dim(B) = \mathrm{tr.deg}(\mathbf{k}(B)/k)$, where $\mathbf{k}(B)$ is the fraction field of B;
(2) for any prime ideal $\mathfrak{p}$ of B,
$$\mathrm{height}(\mathfrak{p}) + \dim(B/\mathfrak{p}) = \dim(B).$$

Proposition 2.25. *For an affine variety Y,*
$$\dim(Y) = \dim(\mathcal{O}(Y)).$$

By the previous fact, the latter equals the number of algebraically independent coordinate functions, and we deduce the following.

Proposition 2.26. *We have that $\dim(\mathbb{A}^n) = n$.*

Proposition 2.27. *Let Y be an affine variety.*

(1) *If Y is irreducible and Z is a proper closed subset of Y, then $\dim(Z) < \dim(Y)$.*
(2) *If $f \in \mathcal{O}(Y)$ is not a zero divisor nor a unit, then $\dim(V(f) \cap Y) = \dim(Y) - 1$.*

Example 2.28.

(1) Let $X, Y \subseteq \mathbb{A}^2$ be two irreducible plane curves. Then $\dim(X \cap Y) < \dim(X) = 1$, so $X \cap Y$ is of dimension 0 and thus it is a finite set.
(2) A classification of irreducible closed subsets of $\mathbb{A}^2$.
   - If $\dim(Y) = 2 = \dim(\mathbb{A}^2)$, then by Proposition 2.27, $Y = \mathbb{A}^2$.
   - If $\dim(Y) = 1$, then $Y \neq \mathbb{A}^2$ so the non-zero ideal $I(Y)$ is prime and thus contains a non-zero irreducible polynomial f.
     Since $Y \supseteq V(f)$ and $\dim(V(f)) = 1$, it must be that $Y = V(f)$.
   - If $\dim(Y) = 0$, then Y is a point.

2.2. Morphisms

Definition 2.29. Let $X \subseteq \mathbb{A}^n$ and $Y \subseteq \mathbb{A}^m$ be affine varieties. A *morphism*
$$\varphi : X \to Y$$
is a map such that there exist polynomials $f_1, \ldots, f_m \in k[x_1, \ldots, x_n]$ with
$$\varphi(P) = (f_1(a_1, \ldots, a_n), \ldots, f_m(a_1, \ldots, a_n)),$$
for every $P = (a_1, \ldots, a_n) \in X$.

Remark 2.30. Morphisms are continuous in the Zariski topology.

A morphism $\varphi : X \to Y$ defines a k-homomorphism
$$\tilde{\varphi} : \mathcal{O}(Y) \to \mathcal{O}(X), \quad \tilde{\varphi}(g) = g \circ \varphi,$$
when $g \in \mathcal{O}(Y)$ is identified with a function $Y \to k$.

A k-homomorphism $\psi : \mathscr{O}(Y) \to \mathscr{O}(X)$ defines a morphism
$$^a\psi : X \to Y.$$
Identify X with $X(k) = \mathrm{Hom}(\mathscr{O}(X), k)$ and Y by $Y(k)$. Then
$$^a\psi(\bar{x}) = \bar{x} \circ \psi.$$

Proposition 2.31. *With the above notation,*
$$^a(\tilde{\varphi}) = \varphi \quad \text{and} \quad \widetilde{(^a\psi)} = \psi.$$

Corollary 2.32. *The functor*
$$X \longmapsto \mathscr{O}(X)$$
defines an arrow-reversing equivalence of categories between the category of affine varieties over k and the category of finitely generated reduced k-algebras.

Indeed, the 'inverse' functor is $A \mapsto \mathrm{Specm}(A)$. For $\psi : B \to A$, $\mathrm{Specm}(\psi) = {}^a\psi : \mathrm{Specm}(A) \to \mathrm{Specm}(B)$, ${}^a\psi(\mathfrak{m}) = \psi^{-1}(\mathfrak{m})$, where $\mathfrak{m}$ is a maximal ideal in A.

In particular, affine varieties X and Y are isomorphic if and only if their coordinate rings $\mathscr{O}(X)$ and $\mathscr{O}(Y)$ are isomorphic as k-algebras.

Thus we have an invaluable *translation mechanism between geometry and algebra:* every time we have a morphism $X \to Y$, we should be thinking that it comes from a morphism $\mathscr{O}(X) \leftarrow \mathscr{O}(Y)$, and vice versa, every time we have a morphism $A \leftarrow B$, we should be thinking of a morphism $\mathrm{Specm}(A) \to \mathrm{Specm}(B)$.

The *methodology of algebraic geometry* adheres to the following principle (to some extent inspired by physics, where one often studies a system X by considering certain 'observable' functions on X).

Study the geometry of X via tools of commutative algebra on $\mathscr{O}(X)$.

Example 2.33.

(1) Let $X = \mathbb{A}^1$ and $Y = V(x^3 - y^2) \subseteq \mathbb{A}^2$, and let
$$\varphi : X \to Y, \quad \text{defined by } t \mapsto (t^2, t^3).$$
Then φ is a morphism which is bijective and bicontinuous (a homeomorphism in Zariski topology), but φ is *not* an isomorphism.

(2) Let $\mathrm{char}(k) = p > 0$. The *Frobenius morphism*
$$\varphi : \mathbb{A}^1 \to \mathbb{A}^1, \quad t \mapsto t^p$$
is a bijective and bicontinuous morphism, but it is not an isomorphism.

2.3. Sheaves

Definition 2.34. Let X be a topological space. A *presheaf* $\mathscr{F}$ of abelian groups on X consists of the following data:

- for every open set $U \subseteq X$, we have an abelian group $\mathscr{F}(U)$;
- for every inclusion $V \xhookrightarrow{i} U$ of open subsets of X, we have a morphism of abelian groups $\rho_{UV} = \mathscr{F}(i) : \mathscr{F}(U) \to \mathscr{F}(V)$, so that

(1) $\rho_{UU} = \mathscr{F}(\mathrm{id} : U \to U) = \mathrm{id} : \mathscr{F}(U) \to \mathscr{F}(U)$;
(2) if $W \xhookrightarrow{j} V \xhookrightarrow{i} U$, then $\mathscr{F}(i \circ j) = \mathscr{F}(j) \circ \mathscr{F}(i)$, i.e., $\rho_{UW} = \rho_{VW} \circ \rho_{UV}$.

In categorical terms, these axioms state that a presheaf $\mathscr{F}$ on a topological space X is nothing other than a *contravariant functor* from the category **Top**(X) of open subsets with inclusions to the category of abelian groups,

$$\mathscr{F} : \mathbf{Top}(X)^{op} \to \mathbf{Ab}.$$

For $s \in \mathscr{F}(U)$ and $V \subseteq U$, we write $s\!\upharpoonright_V = \rho_{UV}(s)$ and we refer to ρ_{UV} as *restrictions*. We can write (2) above as $s\!\upharpoonright_W = (s\!\upharpoonright_V)\!\upharpoonright_W$.

Elements of $\mathscr{F}(U)$ are called *sections* of $\mathscr{F}$ over U, and we sometimes write $\mathscr{F}(U) = \Gamma(U, \mathscr{F})$, where Γ symbolises 'taking sections'.

Definition 2.35. If $P \in X$, the *stalk* $\mathscr{F}_P$ of $\mathscr{F}$ at P is the direct limit of the groups $\mathscr{F}(U)$, where U ranges over the open neighbourhoods of P (via the restriction maps).

Let us define the relation $\sim$ on pairs (U, s), where U is an open neighbourhood of P, and $s \in \mathscr{F}(U)$. We say that $(U_1, s_1) \sim (U_2, s_2)$ if there is an open neighbourhood W of P with $W \subseteq U_1 \cap U_2$ such that $s_1\!\upharpoonright_W = s_2\!\upharpoonright_W$.

Then $\mathscr{F}_P$ equals the set of $\sim$-equivalence classes, which can be thought of as 'germs' of sections at P.

Definition 2.36. A presheaf $\mathscr{F}$ on a topological space X is a *sheaf* provided:

(3) if $\{U_i\}$ is an open covering of U, and $s, t \in \mathscr{F}(U)$ are such that $s\!\upharpoonright_{U_i} = t\!\upharpoonright_{U_i}$ for all i, then $s = t$;
(4) if $\{U_i\}$ is an open covering of U, and $s_i \in \mathscr{F}(U_i)$ are such that for each i, j, $s_i\!\upharpoonright_{U_i \cap U_j} = s_j\!\upharpoonright_{U_i \cap U_j}$, then there exists an $s \in \mathscr{F}(U)$ such that $s\!\upharpoonright_{U_i} = s_i$ (note that such an s is unique by (3)).

Figuratively speaking, sheaves have a *unique gluing property*.

Example 2.37.

- Sheaf $\mathscr{F}$ of continuous $\mathbb{R}$-valued functions on a topological space X:
  - $\mathscr{F}(U)$ is the set of continuous functions $U \to \mathbb{R}$,
  - for $V \subseteq U$, let $\rho_{UV} : \mathscr{F}(U) \to \mathscr{F}(V)$, $\rho_{UV}(f) = f\restriction_V$.
- Sheaf of differentiable functions on a differentiable manifold.
- Sheaf of holomorphic functions on a complex manifold.
- Constant presheaf: fix an abelian group Λ and let $\mathscr{F}(U) = \Lambda$ for all U. This is not a sheaf, and its *associated sheaf* satisfies
$$\mathscr{F}^+(U) = \Lambda^{\pi_0(U)},$$
where $\pi_0(U)$ is the set of connected components of U (provided X is locally connected).

Definition 2.38. Let $\mathscr{F}$ and $\mathscr{G}$ be presheaves of abelian groups on X. A *morphism* $\varphi : \mathscr{F} \to \mathscr{G}$ consists of the following data:

- for each U open in X, we have a morphism
$$\varphi(U) : \mathscr{F}(U) \to \mathscr{G}(U);$$
- for each inclusion $V \overset{i}{\hookrightarrow} U$, we have a diagram

$$\begin{array}{ccc} \mathscr{F}(U) & \xrightarrow{\varphi(U)} & \mathscr{G}(U) \\ {\scriptstyle \mathscr{F}(i)}\downarrow & & \downarrow {\scriptstyle \mathscr{G}(i)} \\ \mathscr{F}(V) & \xrightarrow{\varphi(V)} & \mathscr{G}(V) \end{array}$$

In categorical terms, if $\mathscr{F}$ and $\mathscr{G}$ are considered as functors $\mathbf{Top}(X)^{op} \to \mathbf{Ab}$, a morphism $\varphi : \mathscr{F} \to \mathscr{G}$ is nothing other than a *natural transformation* between these functors.

Definition 2.39. Let A be a commutative ring with 1, and let $S \ni 1$ be a multiplicatively closed subset of A. We define a relation $\equiv$ on $A \times S$:

$$(a_1, s_1) \equiv (a_2, s_2) \quad \text{if } (a_1 s_2 - a_2 s_1)s = 0 \text{ for some } s \in S.$$

Then $\equiv$ is an equivalence relation and the *ring of fractions* $S^{-1}A = A \times S/\equiv$ has the following structure (write a/s for the class of (a, s)):

$$(a_1/s_1) + (a_2/s_2) = (a_1 s_2 + a_2 s_1)/s_1 s_2,$$
$$(a_1/s_1)(a_2/s_2) = (a_1 a_2/s_1 s_2).$$

We have a morphism $A \to S^{-1}A$, $a \mapsto a/1$.

Example 2.40.

- If A is a domain, $S = A \setminus \{0\}$, then $S^{-1}A$ is the field of fractions of A.
- If $\mathfrak{p}$ is a prime ideal in A, then $S = A \setminus \mathfrak{p}$ is multiplicative and $S^{-1}A$ is denoted $A_\mathfrak{p}$ and called the *localisation* of A at $\mathfrak{p}$. Note that $A_\mathfrak{p}$ is indeed a *local ring*, i.e., it has a unique maximal ideal.
- Let $f \in A$, $S = \{f^n : n \geq 0\}$. Write $A_f = S^{-1}A$.
- We have that $S^{-1}A = 0$ if and only if $0 \in S$.

Remark 2.41. Let X be an affine variety and $g, h \in \mathscr{O}(X)$. Then

$$P \mapsto \frac{g(P)}{h(P)}$$

is a well-defined function $D(h) \to k$.

We would like to consider functions defined on open subsets of X which are locally of this form.

Definition 2.42. Let U be an open subset of an affine variety X.

- A function $f : U \to k$ is *regular* if for every $P \in U$, there exist $g, h \in \mathscr{O}(X)$ with $h(P) \neq 0$, and a neighbourhood V of P such that the functions f and g/h agree on V.
- The set of all regular functions on U is denoted $\mathscr{O}_X(U)$.

Proposition 2.43. *The assignment $U \mapsto \mathscr{O}_X(U)$ defines a sheaf of k-algebras on X.*

The sheaf $\mathscr{O}_X$ is called the *structure sheaf* of X.

Proposition 2.44. *Let X be an affine variety and let $A = \mathscr{O}(X)$ be its coordinate ring. Then we have the following conditions.*

- *For any $P \in X$, the stalk $\mathscr{O}_{X,P}$ is isomorphic to $A_{\mathfrak{m}_P}$, where the maximal ideal $\mathfrak{m}_P = \{f \in A : f(P) = 0\}$ is the image of $I(P)$ in A.*
- *For any $f \in A$,*

$$\mathscr{O}_X(D(f)) \simeq A_f.$$

- *In particular,*

$$\mathscr{O}_X(X) = A.$$

The last claim shows that our notation for the coordinate ring is justified.

2.4. *Schemes*

Let A be a commutative ring with identity.

Definition 2.45. Let $\mathrm{Spec}(A)$ be the set of all prime ideals in A.

Our goal is to turn $X = \mathrm{Spec}(A)$ into a topological space and equip it with a sheaf of rings, i.e., make it into a *ringed space*. We shall use the following notation:

- write $x \in X$ for a point, and j_x for the corresponding prime ideal in A;
- $A_x = A_{\mathrm{j}_x}$, the local ring at x;
- $\mathfrak{m}_x = \mathrm{j}_x A_{\mathrm{j}_x}$, the maximal ideal of A_x;
- $\mathbf{k}(x) = A_x/\mathfrak{m}_x$, the residue field at x, naturally isomorphic to the fraction field of the domain A/j_x;
- for $f \in A$, write $f(x)$ for the class of f mod j_x in $\mathbf{k}(x)$. Then '$f(x) = 0$' if and only if $f \in \mathrm{j}_x$.

Example 2.46.

(1) For a field F, $\mathrm{Spec}(F) = \{0\}$, $\mathbf{k}(0) = F$.
(2) Let $\mathbb{Z}_p$ be the ring of p-adic integers. $\mathrm{Spec}(\mathbb{Z}_p) = \{0, (p)\}$, and $\mathbf{k}(0) = \mathbb{Q}_p$, $\mathbf{k}((p)) = \mathbb{F}_p$. This generalises to an arbitrary DVR (Definition 3.11).
(3) $\mathrm{Spec}(\mathbb{Z}) = \{0\} \cup \{(p) : p \text{ prime}\}$. Residue fields are $\mathbf{k}(0) = \mathbb{Q}$, $\mathbf{k}((p)) = \mathbb{F}_p$. For $f \in \mathbb{Z}$, $f(0) = f/1 \in \mathbb{Q}$, and $f(p) = f \mod p \in \mathbb{F}_p$.
(4) For an algebraically closed field k, let $A = k[x, y]$. Then, by Example 2.28,

$$\mathrm{Spec}(A) = \{0\} \cup \{(x-a, y-b) : a, b \in k\} \cup \{(g) : g \in A \text{ irreducible}\}.$$

Note that $\mathbf{k}(0) = k(x, y)$, $\mathbf{k}((x-a, y-b)) = k$, and $\mathbf{k}((g))$ is the fraction field of the domain $A/(g)$. For $f \in A$, $f(0) = f/1 \in k(x, y)$, $f((x-a, x-b)) = f(a, b) \in k$, $f((g)) = (f+(g))/1 \in \mathbf{k}(g)$.

Definition 2.47. For $f \in A$ and $E \subseteq A$, let

$$V(f) = \{x \in X : f \in \mathrm{j}_x\}, \quad \text{i.e., the set of } x \text{ with } f(x) = 0;$$

$$D(f) = X \setminus V(f);$$

$$V(E) = \bigcap_{f \in E} V(f) = \{x \in X : E \subseteq \mathrm{j}_x\}.$$

The operation V has the expected properties (as in Proposition 2.3), and the sets of the form $V(E)$ are closed sets for the *Zariski topology* on X.

Definition 2.48. For an arbitrary subset $Y \subseteq X$, the *ideal* of Y is
$$j(Y) = \bigcap_{x \in Y} j_x \quad \text{i.e., the set of } f \in A \text{ with } f(x) = 0 \text{ for } x \in Y.$$

Remark 2.49. For any $E \subseteq A$,
$$\sqrt{E} = \bigcap_{x \in V(E)} j_x.$$

The operation j has the expected properties (as in Proposition 2.7) and here the proof is trivial, there is no need for Nullstellensatz.

Definition 2.50. Let $\varphi : X \to Y$ be a continuous map of topological spaces and let $\mathscr{F}$ be a presheaf on X. The *direct image* $\varphi_* \mathscr{F}$ is a presheaf on Y defined by
$$\varphi_* \mathscr{F}(U) = \mathscr{F}(\varphi^{-1} U).$$

Lemma 2.51. *If $\mathscr{F}$ is a sheaf, so is $\varphi_* \mathscr{F}$.*

Definition 2.52.

- A *ringed space* $(X, \mathscr{O}_X)$ consists of a topological space X and a sheaf of rings $\mathscr{O}_X$ on X, called the *structure sheaf*.
- A *locally ringed space* is a ringed space $(X, \mathscr{O}_X)$ such that every stalk $\mathscr{O}_{X,x}$ is a local ring, $x \in X$.
- A *morphism of ringed spaces* $(X, \mathscr{O}_X) \to (Y, \mathscr{O}_Y)$ is a pair (φ, θ), where $\varphi : X \to Y$ is a continuous map, and
$$\theta : \mathscr{O}_Y \to \varphi_* \mathscr{O}_X$$
is a map of sheaves.
- (φ, θ) is a *morphism of locally ringed spaces*, if, additionally, each induced map of stalks
$$\theta_x^\sharp : \mathscr{O}_{Y, \varphi(x)} \to \mathscr{O}_{X,x}$$
is a *local homomorphism* of local rings.

Lemma 2.53. *There exists a unique sheaf $\mathscr{O}_X$ on $X = \mathrm{Spec}(A)$ satisfying*
$$\mathscr{O}_X(D(f)) \simeq A_f \quad \text{for } f \in A.$$

Its stalks are
$$\mathscr{O}_{X,x} \simeq A_x \ (= A_{j_x}).$$

Definition 2.54. By $\mathrm{Spec}(A)$ we shall mean the locally ringed space
$$(\mathrm{Spec}(A), \mathscr{O}_{\mathrm{Spec}(A)}).$$

Definition 2.55.

- An *affine scheme* is a ringed space $(X, \mathcal{O}_X)$ which is isomorphic to $\mathrm{Spec}(A)$ for some ring A.
- A *scheme* is a ringed space $(X, \mathcal{O}_X)$ such that every point has an open affine neighbourhood U (i.e., $(U, \mathcal{O}_X\!\restriction_U)$ is an affine scheme).
- A *morphism* $(X, \mathcal{O}_X) \to (Y, \mathcal{O}_Y)$ is a morphism of locally ringed spaces.

Definition 2.56. A ring homomorphism $\varphi : B \to A$ gives rise to a morphism of affine schemes $X = \mathrm{Spec}(A)$ and $Y = \mathrm{Spec}(B)$:
$$({}^a\varphi, \tilde{\varphi}) : (\mathrm{Spec}(A), \mathcal{O}_{\mathrm{Spec}(A)}) \to (\mathrm{Spec}(B), \mathcal{O}_{\mathrm{Spec}(B)}), \text{ where}$$

- ${}^a\varphi(x) = y$ if and only if $\mathfrak{j}_y = \varphi^{-1}(\mathfrak{j}_x)$ (i.e., ${}^a\varphi(\mathfrak{p}) = \varphi^{-1}(\mathfrak{p})$);
- $\tilde{\varphi} : \mathcal{O}_Y \to {}^a\varphi_* \mathcal{O}_X$ is characterised by (for $g \in B$):

$$
\begin{array}{ccc}
\mathcal{O}_Y(D(g)) & \xrightarrow{\tilde{\varphi}(D(g))} & \mathcal{O}_X({}^a\varphi^{-1}D(g)) = \mathcal{O}_X(D(\varphi(g))) \\
\| & & \| \\
B_g & \xrightarrow{b/g^n \,\longmapsto\, \varphi(b)/\varphi(g)^n} & A_{\varphi(g)}
\end{array}
$$

It turns out that *every* morphism of affine schemes is induced by a ring homomorphism.

Proposition 2.57. *There is a canonical isomorphism*
$$\mathrm{Hom}(\mathrm{Spec}(A), \mathrm{Spec}(B)) \simeq \mathrm{Hom}(B, A).$$

Corollary 2.58. *The functors*
$$A \longmapsto \mathrm{Spec}(A)$$
$$\mathcal{O}_X(X) \longleftarrow\!\shortmid X$$
define an arrow-reversing equivalence of categories between the category of commutative rings *and the* category of affine schemes.

More generally, we have the following proposition.

Proposition 2.59. *Let X be an arbitrary scheme, and let A be a ring. There is a canonical isomorphism*
$$\mathrm{Hom}(X, \mathrm{Spec}(A)) \simeq \mathrm{Hom}(A, \Gamma(X)),$$
where $\Gamma(X) = \mathcal{O}_X(X)$ is the 'global sections' functor.

Proposition 2.60. *Let X_1 and X_2 be schemes. There exists a scheme $X_1 \amalg X_2$, called the* sum *of X_1 and X_2, together with morphisms $X_i \to X_1 \amalg X_2$ such that for every scheme Z*

$$\mathrm{Hom}(X_1 \amalg X_2, Z) \simeq \mathrm{Hom}(X_1, Z) \times \mathrm{Hom}(X_2, Z),$$

i.e., every solid commutative diagram

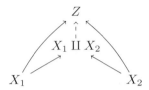

can be completed by a unique dashed morphism.

Proof. We reduce to affine schemes $X_i = \mathrm{Spec}(A_i)$. Then $X_1 \amalg X_2 = \mathrm{Spec}(A_1 \times A_2)$, and the underlying topological space of $X_1 \amalg X_2$ is the disjoint union of the X_i. □

Definition 2.61. Let us fix a scheme S.

- An *S-scheme*, or a *scheme over S* is a morphism $X \to S$.
- A *morphism* of S-schemes is a commutative diagram

Example 2.62.

- Let k be a field (or even a ring) and $S = \mathrm{Spec}(k)$. The category of affine S-schemes is anti-equivalent to the category of k-algebras.
- If k is algebraically closed, and we consider only reduced finitely generated k-algebras, the resulting category is essentially the category of affine algebraic varieties over k.

Proposition 2.63. *Let X_1 and X_2 be schemes over S. There exists a scheme $X_1 \times_S X_2$, called the* (fibre) product *of X_1 and X_2 over S, together with S-morphisms $X_1 \times_S X_2 \to X_i$ such that for every S-scheme Z,*

$$\mathrm{Hom}_S(Z, X_1 \times_S X_2) \simeq \mathrm{Hom}_S(Z_1, X_1) \times \mathrm{Hom}_S(Z, X_2),$$

i.e., every solid commutative diagram

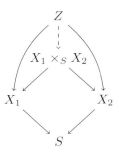

can be completed by a unique dashed morphism.

Proof. We reduce to affine schemes $X_i = \text{Spec}(A_i)$ over $S = \text{Spec}(R)$. Then A_i are R-algebras and $X_1 \times_S X_2 = \text{Spec}(A_1 \otimes_R A_2)$. $\square$

Definition 2.64. Let X and T be schemes. The set of *T-valued points of X* is the set
$$X(T) = \text{Hom}(T, X).$$

In a relative setting, suppose X, T are S-schemes. The set of *T-valued points of X over S* is the set
$$X(T)_S = \text{Hom}_S(T, X).$$

Example 2.65. This notation is most commonly used as follows. Consider:

- a system of polynomial equations $f_i = 0$, $i = 1, \ldots, m$ defined over a field k, i.e., $f_i \in k[x_1, \ldots, x_n]$;
- $A = k[x_1, \ldots, x_n]/(f_1, \ldots, f_n)$;
- a field extension $K \supseteq k$.

The associated scheme is $X = \text{Spec}(A)$. Then
$$X(K)_k = \text{Hom}_{\text{Spec}(k)}(\text{Spec}(K), X) \simeq \text{Hom}_k(A, K)$$
$$\simeq \{\bar{a} \in K^n : f_i(\bar{a}) = 0 \text{ for all } i\}.$$

When k is algebraically closed, the set $X(k) := X(k)_k \subseteq k^n$ is what we called an *affine variety* $V(\{f_i\})$ at the start. The scheme X contains a whole lot more information.

Example 2.66. Suppose S is a scheme over a field k, and let $X \xrightarrow{f} S$, $Y \xrightarrow{g} S$ be two schemes over S (in particular, over k). Then

$$(X \times_S Y)(k) = X(k) \times_{S(k)} Y(k)$$
$$= \{(\bar{x}, \bar{y}) : \bar{x} \in X(k), \bar{y} \in Y(k), \ f(\bar{x}) = g(\bar{y})\}.$$

Example 2.67. Zariski topology of the product is *not* the product topology, as shown in the following example. For a field k and $\mathbb{A}^1 = \mathbb{A}^1_k$, we have

$$\mathbb{A}^1 \times \mathbb{A}^1 = \mathbb{A}^1 \times_{\mathrm{Spec}(k)} \mathbb{A}^1$$
$$= \mathrm{Spec}(k[x_1] \otimes_k k[x_2]) \simeq \mathrm{Spec}(k[x_1, x_2]) = \mathbb{A}^2.$$

The set of k-points $\mathbb{A}^2(k)$ is the cartesian product $\mathbb{A}^1(k) \times \mathbb{A}^1(k)$. However, as a scheme, $\mathbb{A}^2$ has more points than the cartesian square of the set of points of $\mathbb{A}^1$.

Definition 2.68. Let $\varphi : X \to S$ be a morphism, and let $s \in S$. There exists a natural morphism $\mathrm{Spec}(\mathbf{k}(s)) \to S$. The *fibre* of φ over s is

$$X_s = X \times_S \mathrm{Spec}(\mathbf{k}(s)).$$

Remark 2.69. The fibre X_s should be thought of as $\varphi^{-1}(s)$, except that the above definition gives it a structure of a $\mathbf{k}(s)$-scheme.

Example 2.70. Consider $R = k[z] \to A = k[x, y, z]/(y^2 - x(x-1)(x-z))$ and the corresponding morphism $\varphi : X = \mathrm{Spec}(A) \to S = \mathrm{Spec}(R)$. Then, for $s \in S$ corresponding to the ideal $(z - \lambda)$, $\lambda \in k$,

$$X_s = X_\lambda = \mathrm{Spec}(k[x, y]/(y^2 - x(x-1)(x-\lambda)))$$

so we can consider φ as a *family* of curves X_s with parameters s from S.

Example 2.71. Consider $\mathbb{Z} \to A = \mathbb{Z}[x, y]/(y^2 - x^3 - x - 1)$ and the corresponding morphism $\varphi : X = \mathrm{Spec}(A) \to S = \mathrm{Spec}(\mathbb{Z})$. Then, for $\mathfrak{p} \in S$ corresponding to the ideal (p) for a prime integer p,

$$X_\mathfrak{p} = X_p = \mathrm{Spec}(\mathbb{F}_p[x, y]/(y^2 - x^3 - x - 1)),$$

as a scheme over $\mathbb{F}_p = \mathbf{k}(\mathfrak{p})$, considered as a *reduction* of X modulo p.

2.5. *Projective varieties*

Fix an algebraically closed field k.

Definition 2.72. The *projective n-space* over k is the set $\mathbb{P}^n_k$ of equivalence classes
$$[a_0 : a_1 : \cdots : a_n]$$
of $(n+1)$-tuples $(a_0, a_1, \ldots, a_n)$ of elements of k, not all zero, under the equivalence relation
$$(a_0, \ldots, a_n) \sim (\lambda a_0, \ldots, \lambda a_n),$$
for all $\lambda \in k \setminus \{0\}$.

Example 2.73 (Projective line as a compactification of $\mathbb{A}^1$). The understanding of $\mathbb{P}^1$ can be enhanced by the picture of the stereographic projection,

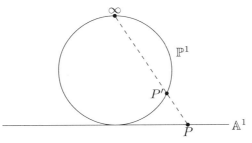

allowing us to think of $\mathbb{P}^1$ as the set of lines through a fixed point. Note that an equation $ax + by + c = 0$ describes a line if not both a, b are 0, and what really matters is the ratio $[a : b]$.

Let $S = k[x_0, \ldots, x_n]$. For an arbitrary $f \in S$ and $P = [a_0 : \cdots : a_n] \in \mathbb{P}^n$, the expression $f(P)$ *does not* make sense.

On the other hand, for a *homogeneous* polynomial $f \in S$ of degree d,
$$f(\lambda a_0, \ldots, \lambda a_n) = \lambda^d f(a_0, \ldots, a_n),$$
so it *does* make sense to consider whether $f(P) = 0$ or $f(P) \neq 0$.

It is therefore beneficial to consider S as a *graded ring*
$$S = \bigoplus_{d \geq 0} S_d,$$
where S_d is the abelian group consisting of degree d homogeneous polynomials.

Definition 2.74. Let $T \subseteq S$ be a set of *homogeneous* polynomials. Let
$$V(T) = \{P \in \mathbb{P}^n : f(P) = 0 \text{ for all } f \in T\}.$$
For a homogeneous polynomial $f \in S$, we write
$$D(f) = \mathbb{P}^n \setminus V(f).$$
The operation V has the expected properties and it gives rise to the *Zariski topology* on $\mathbb{P}^n$.

Definition 2.75. A *projective algebraic variety* is a subset of $\mathbb{P}^n$ of the form $V(T)$, together with the induced Zariski topology.

Remark 2.76. Let
$$U_i = D(x_i) = \{[a_0 : \cdots : a_n] \in \mathbb{P}^n : a_i \neq 0\} \subseteq \mathbb{P}^n, \quad i = 0, \ldots, n.$$
The maps $\varphi_i : U_i \to \mathbb{A}^n$ defined by
$$\varphi_i([a_0 : \cdots : a_n]) = \left(\frac{a_0}{a_i}, \ldots, \frac{a_{i-1}}{a_i}, \frac{a_{i+1}}{a_i}, \ldots, \frac{a_n}{a_i}\right)$$
are all homeomorphisms. Thus we can cover $\mathbb{P}^n$ by $n+1$ affine open subsets.

Example 2.77 (From an affine to a projective curve). Start with your favourite plane curve, e.g., $X = V(y^2 - x^3 - x - 1)$. Substitute $y \leftarrow y/z$, $x \leftarrow x/z$ to get
$$\frac{y^2}{z^2} = \frac{x^3}{z^3} + \frac{x}{z} + 1.$$
Clearing the denominators yields
$$y^2 z = x^3 + xz^2 + z^3.$$
This is a homogeneous equation of a projective curve $\tilde{X}$ in $\mathbb{P}^2$, and $\tilde{X} \cap D(z) \simeq X$.

Definition 2.78. We say that S is a *graded ring* if

- $S = \bigoplus_{d \geq 0} S_d$, where S_d are abelian subgroups;
- $S_d \cdot S_e \subseteq S_{d+e}$.

An element $f \in S_d$ is *homogeneous* of degree d. An ideal $\mathfrak{a}$ in S is *homogeneous* if it is generated by homogeneous elements, i.e.,
$$\mathfrak{a} = \bigoplus_{d \geq 0} (\mathfrak{a} \cap S_d).$$

Definition 2.79. For a graded ring S, let
$$S_+ = \bigoplus_{d>0} S_d \trianglelefteq S, \quad \text{and}$$
$$\operatorname{Proj}(S) = \{\mathfrak{p} \trianglelefteq S : \mathfrak{p} \text{ prime, and } S_+ \not\subseteq \mathfrak{p}\}.$$

For a homogeneous ideal $\mathfrak{a}$, and a homogeneous element $f \in S$, let
$$V_+(\mathfrak{a}) = \{\mathfrak{p} \in \operatorname{Proj}(S) : \mathfrak{p} \supseteq \mathfrak{a}\} \quad \text{and} \quad D_+(f) = \operatorname{Proj}(S) \setminus V_+(f).$$

As expected, V_+ makes $\operatorname{Proj}(S)$ into a topological space.

We would like to define the structure sheaf on $\operatorname{Proj}(S)$ next. For $\mathfrak{p} \in \operatorname{Proj}(S)$, write $S_{(\mathfrak{p})}$ for the ring of degree zero elements in $T^{-1}S$, where T is the multiplicative set of homogeneous elements in $S \setminus \mathfrak{p}$.

If $a, f \in S$ are homogeneous of the same degree, then the function $P \mapsto a(P)/f(P)$ makes sense on $D_+(f)$.

Definition 2.80. For U open in $\operatorname{Proj}(S)$, we define $\mathscr{O}(U)$ as the set of maps $s : U \to \coprod_{\mathfrak{p} \in U} S_{(\mathfrak{p})}$ such that, for each $\mathfrak{p} \in U$, $s(\mathfrak{p}) \in S_{(\mathfrak{p})}$, and for each $\mathfrak{p}$ there is a neighbourhood $V \ni \mathfrak{p}$, $V \subseteq U$ and homogeneous elements a, f of the same degree such that for all $\mathfrak{q} \in V$, $f \notin \mathfrak{q}$, and $s(\mathfrak{q}) = a/f$ in $S_{(\mathfrak{q})}$.

Proposition 2.81.

(1) *For $\mathfrak{p} \in \operatorname{Proj}(S)$, the stalk $\mathscr{O}_\mathfrak{p}$ is isomorphic to $S_{(\mathfrak{p})}$.*
(2) *The sets $D_+(f)$, for $f \in S$ homogeneous, cover $\operatorname{Proj}(S)$, and*
$$(D_+(f), \mathscr{O}\restriction_{D_+(f)}) \simeq \operatorname{Spec}(S_{(f)}),$$
where $S_{(f)}$ is the subring of elements of degree 0 in S_f.
(3) $\operatorname{Proj}(S)$ *is a scheme.*

Thus we obtained an example of a scheme which is not affine.

Remark 2.82. The property (2) above shows that
$$\mathscr{O}(\operatorname{Proj}(S)) = S_0,$$
so the only global regular functions on $\mathbb{P}^n = \operatorname{Proj}(k[x_0, \ldots, x_n])$ are constant functions, since $k[x_0, \ldots, x_n]_0 = k$. The same statement holds for projective varieties.

2.6. First properties of schemes

Definition 2.83. A scheme X is said to be

- *connected/quasi-compact/irreducible*, if it is so topologically;
- *reduced*, if for every open U, $\mathscr{O}_X(U)$ has no nilpotents;
- *integral*, if every $\mathscr{O}_X(U)$ is an integral domain.

Lemma 2.84. *A scheme is integral if and only if it is reduced and irreducible.*

Definition 2.85 (Finiteness properties). Let X and Y be schemes.

- X is *noetherian* if it can be covered by finitely many open affine $\operatorname{Spec}(A_i)$ with each A_i a noetherian ring;
- $\varphi : X \to Y$ is *of finite type* if there exists a covering of Y by open affines $V_i = \operatorname{Spec}(B_i)$ such that for each i, $\varphi^{-1}(V_i)$ can be covered by finitely many open affines $U_{ij} = \operatorname{Spec}(A_{ij})$ where each A_{ij} is a finitely generated B_i-algebra;
- $\varphi : X \to Y$ is *finite* if Y can be covered by open affines $V_i = \operatorname{Spec}(B_i)$ such that for each i, $\varphi^{-1}(V_i) = \operatorname{Spec}(A_i)$ with A_i is a B_i-algebra which is a finitely generated B_i-module.

Definition 2.86. Let $f : X \to Y$ be a morphism. We say that f is

- *separated*, if the diagonal Δ is closed in $X \times_Y X$;
- *closed*, if the image of any closed subset is closed;
- *universally closed*, if every base change of it is closed; for every morphism $Y' \to Y$, the corresponding morphism $X \times_Y Y' \to Y'$ is closed;
- *proper*, if it is separated, of finite type and universally closed.

Hereafter, a *variety* is a reduced separated scheme of finite type over a field, and a *curve* is a variety of dimension 1.

Example 2.87. Finite morphisms are proper.

Proposition 2.88. *Projective varieties are proper (over k).*

Indeed, projective varieties are algebraic analogues of compact manifolds.

Example 2.89. Let $Z = V(xy - 1)$, $X = \mathbb{A}^1$ and let $\pi : Z \to X$ be the projection $(x, y) \mapsto x$. The image $\pi(Z) = \mathbb{A}^1 \setminus \{0\}$ is not closed.

Theorem 2.90 (Chevalley). *Let $f : X \to Y$ be a morphism of schemes of finite type. Then the image of a constructible set is a constructible set (i.e., a Boolean combination of closed subsets).*

3. Local Properties

3.1. Non-singular schemes

Suppose we have a point $P = (a, b)$ on a plane curve X defined by equation $f(x, y) = 0$. In analysis, the *tangent* to X at P is the line

$$\frac{\partial f}{\partial x}(P)(x - a) + \frac{\partial f}{\partial y}(P)(y - b) = 0.$$

The partial derivatives of a polynomial make sense over *any* field or ring. Note, in order for the 'tangent line' to be defined, we need at least one of $\frac{\partial f}{\partial x}(P)$, $\frac{\partial f}{\partial y}(P)$ to be non-zero. Otherwise, P can be considered 'singular'.

Example 3.1. The curve $y^2 = x^3$ has a singular point $(0, 0)$.

Definition 3.2. Let $X \subseteq \mathbb{A}^n$ be an irreducible affine variety, $I = I(X)$, $P = (a_1, \ldots, a_n) \in X$. The *tangent space* $T_P(X)$ to X at P is the solution set of all linear equations

$$\sum_{i=1}^{n} \frac{\partial f}{\partial x_i}(P)(x_i - a_i) = 0, \quad f \in I.$$

It is enough to take f from a generating set of I.

We say that P is *non-singular* on X if

$$\dim_k T_P(X) = \dim X.$$

Definition 3.3. Let A be a ring, B an A-algebra, and M a module over B. An A-*derivation* of B into M is a map $d : B \to M$ satisfying

(1) d is additive;
(2) $d(bb') = b d(b') + b' d(b)$;
(3) $d(a) = 0$ for $a \in A$.

Definition 3.4. The *module of relative differential forms* of B over A is a B-module $\Omega_{B/A}$ together with an A-derivation $d : B \to \Omega_{B/A}$ such that for any A-derivation $d' : B \to M$, there exists a unique B-module homomorphism $f : \Omega_{B/A} \to M$ such that $d' = f \circ d$, as in the following diagram.

$$\begin{array}{ccc} B & \xrightarrow{d} & \Omega_{B/A} \\ & \searrow{\scriptstyle d'} & \downarrow{\scriptstyle \exists! f} \\ & & M \end{array}$$

The module $\Omega_{B/A}$ is obtained as a quotient of the free B-module generated by symbols $\{db : b \in B\}$ by the submodule generated by elements:

(1) $d(bb') - bd(b') - b'd(b)$, for $b, b' \in B$;
(2) da, for $a \in A$.

The 'universal' derivation is simply

$$d : b \longmapsto \text{(the coset of) } db.$$

Lemma 3.5. *Let X be an affine variety over an algebraically closed field k, $P \in X$. Let $\mathfrak{m}_P$ be the maximal ideal of $\mathscr{O}_P$. We have isomorphisms*

$$\mathrm{Der}_k(\mathscr{O}_P, k) \xrightarrow{\sim} \mathrm{Hom}_{k\text{-}linear}(\mathfrak{m}_P/\mathfrak{m}_P^2, k) \xrightarrow{\sim} T_P(X).$$

Hence $\Omega_{\mathscr{O}_P/k} \otimes_{\mathscr{O}_P} k \simeq \mathfrak{m}_P/\mathfrak{m}_P^2$.

We conclude that P is non-singular if and only if $\dim_k(\mathfrak{m}_P/\mathfrak{m}_P^2) = \dim(\mathscr{O}_P)$, or, equivalently, if $\Omega_{\mathscr{O}_P/k}$ is a free $\mathscr{O}_P$-module of rank $\dim(\mathscr{O}_P)$.

Definition 3.6. A noetherian local ring $(R, \mathfrak{m})$ with residue field $k = R/\mathfrak{m}$ is *regular*, if $\dim_k(\mathfrak{m}/\mathfrak{m}^2) = \dim(R)$.

By Nakayama's lemma [1], this is equivalent to $\mathfrak{m}$ having $\dim(R)$ generators.

Definition 3.7. A noetherian scheme X is *regular*, or *non-singular* at x, if $\mathscr{O}_x$ is a regular local ring. The scheme X is *regular/non-singular* if it is so at every point $x \in X$.

Let $\varphi : X \to Y$ be a morphism. There exists a *sheaf of relative differentials* $\Omega_{X/Y}$ on X and a sheaf morphism $d : \mathscr{O}_X \to \Omega_{X/Y}$ such that whenever $U = \mathrm{Spec}(A) \subseteq Y$ and $V = \mathrm{Spec}(B) \subseteq X$ are open affine such that $\varphi(V) \subseteq U$, then $\Omega_{X/Y}(V) = \Omega_{B/A}$.

Proposition 3.8. *Let X be an irreducible scheme of finite type over an algebraically closed field k. Then X is regular if and only if $\Omega_{X/k}$ is a locally free sheaf of rank $\dim(X)$, i.e., every point has an open neighbourhood U such that*

$$\Omega_{X/k}\restriction_U \simeq (\mathscr{O}_X\restriction_U)^{\dim(X)}.$$

The latter property is associated with a notion of *smoothness*, and it coincides with regularity over a perfect base field.

Corollary 3.9. *If X is a variety over a field k of characteristic 0, then there is an open dense subset U of X which is non-singular.*

3.2. *Multiplicities*

Definition 3.10. A *discrete valuation* of a field K is a map $v : K\setminus\{0\} \to \mathbb{Z}$ such that

(1) $v(xy) = v(x) + v(y)$;
(2) $v(x+y) \geq \min(v(x), v(y))$.

Then $R = \{x \in K : v(x) \geq 0\} \cup \{0\}$ is a subring of K, called the *valuation ring*, and $\mathfrak{m} = \{x \in K : v(x) > 0\} \cup \{0\}$ is a maximal ideal in R, and $(R, \mathfrak{m})$ is a local ring.

Definition 3.11. A *discrete valuation ring (DVR)* is an integral domain R which is the valuation ring of some discrete valuation of $\mathrm{Fract}(R)$.

Proposition 3.12. *Let $(R, \mathfrak{m})$ be a noetherian local domain of dimension 1. The following properties are equivalent:*

(1) *R is a DVR;*
(2) *R is integrally closed;*
(3) *R is a regular local ring;*
(4) *$\mathfrak{m}$ is a principal ideal.*

Remark 3.13. Let X be a non-singular curve, $x \in X$. Then $\mathscr{O}_x$ is a regular local ring of dimension 1, and thus a DVR.

A *uniformiser* at x is a generator of $\mathfrak{m}_x$.

Proposition 3.14. *Let R be an integral domain which is not a field. The following statements are equivalent.*

(1) *every nonzero proper ideal factors into primes*;
(2) *R is noetherian, and the localisation at every maximal ideal is a DVR*;
(3) *R is an integrally closed noetherian domain of dimension* 1.

Definition 3.15. A ring is a *Dedekind domain* if it satisfies (any of) the above equivalent conditions.

Remark 3.16. If X is a non-singular affine curve, then $\mathcal{O}(X)$ is a Dedekind domain.

3.3. *Divisors*

Definition 3.17. Let X be a connected non-singular curve over an algebraically closed field k and let X^0 denote the set of closed points of X.

- A *Weil divisor* is an element of the free abelian group $\operatorname{Div} X$ generated by X^0, i.e., it is a formal integer combination of (closed) points of X.
- A divisor $D = \sum_i n_i x_i$ is *effective*, denoted $D \geq 0$, if all $n_i \geq 0$.

Definition 3.18. Let X be a connected non-singular curve over an algebraically closed field k, and let $K = \mathbf{k}(X) = \mathcal{O}_\xi = \varinjlim_{U \text{ open}} \mathcal{O}_X(U)$ be its *function field* (where ξ is the generic point of X), which we think of as the field of 'rational functions' on X.

For $f \in K^\times$, we let the *divisor* (f) *of* f *on* X (also written $\operatorname{div}(f)$) be

$$(f) = \sum_{x \in X^0} v_x(f) \cdot x,$$

where v_x is the valuation in $\mathcal{O}_x$. Such divisors are called *principal*.

Remark 3.19. Note that this is a divisor: if f is represented as $f_U \in \mathcal{O}_X(U)$ on some open U, then (f) is 'supported' on $V(f_U) \cup X \setminus U$, which is a proper closed subset of X and it is thus finite.

Remark 3.20. The assignment $f \mapsto (f)$ is a homomorphism $K^\times \to \operatorname{Div} X$ whose image is the subgroup of principal divisors.

Definition 3.21. For a divisor $D = \sum_i n_i x_i$, we define the *degree* of D as

$$\deg(D) = \sum_i n_i,$$

making deg into a homomorphism $\operatorname{Div} X \to \mathbb{Z}$.

Definition 3.22. Let X be a non-singular curve over k.

- Two divisors $D, D' \in \text{Div}\, X$ are *linearly equivalent*, written $D \sim D'$, if $D - D'$ is a principal divisor.
- The *divisor class group* $\text{Cl}\, X$ is the quotient of $\text{Div}\, X$ by the subgroup of principal divisors.

Definition 3.23. Let $\varphi : X \to Y$ be a morphism of non-singular curves, $y \in Y$ and $x \in X$ with $\pi(x) = y$. The *ramification index* of φ at x is
$$e_x(\varphi) = v_x(\varphi^\sharp t_y),$$
where $\varphi^\sharp$ is the local morphism $\mathcal{O}_y \to \mathcal{O}_x$ induced by φ and t_y is a uniformiser at y, i.e., $\mathfrak{m}_y = (t_y)$.

When φ is finite, we can define a morphism $\varphi^* : \text{Div}\, Y \to \text{Div}\, X$ by extending the rule
$$\varphi^*(y) = \sum_{\varphi(x)=y} e_x(\varphi) \cdot x$$
for prime divisors $y \in Y$ by linearity to $\text{Div}\, Y$.

Theorem 3.24 (Preservation of multiplicity). *Let $\varphi : X \to Y$ be a morphism of non-singular projective curves with $\varphi(X) = Y$, then $\deg \varphi = \deg(\varphi^*(y))$ for any point $y \in Y$.*

The proof reduces to the Chinese Remainder Theorem.

Corollary 3.25. *The degree of a principal divisor on a non-singular projective curve equals 0.*

Proof. Any $f \in \mathbf{k}(X)$ defines a morphism $f : X \to \mathbb{P}^1$. Then $\deg((f)) = \deg(f^*(0)) - \deg(f^*(\infty)) = \deg(f) - \deg(f) = 0$. □

Remark 3.26. Hence $\deg : \text{Cl}(X) \to \mathbb{Z}$ is well defined.

Theorem 3.27 (Bezout). *Let $X \subseteq \mathbb{P}^n$ be a non-singular projective curve, and let $H = V_+(f) \subseteq \mathbb{P}^n$ be the hypersurface defined by a homogeneous polynomial f. Then, writing*
$$X.H = \sum_{x \in X \cap H} i(x; X, H) x := (f),$$
we have that
$$\deg(X.H) = \deg(X) \deg(f),$$
where $\deg(X)$ is the maximal number of points of intersection of X with a hyperplane in $\mathbb{P}^n$ (which does not contain a component of X).

Proof. Let $d = \deg(f)$. For any linear form l, $h = f/l^d \in \mathbf{k}(X)$, so $\deg((f)) = \deg((l^d)) + \deg((h)) = d \deg(l) + 0 = d \deg(X)$. □

Let E be a non-singular projective plane cubic, and pick a point $o \in E$. For points $p, q \in E$, let $p * q$ be the unique point such that, writing L for the line pq and using Bezout, $E.L = p + q + p * q$. We define
$$p \oplus q = o * (p * q).$$

Example 3.28. Let E be given by $y^2 z = x^3 - 2xz^2$, and $o = \infty := [0:1:0]$.

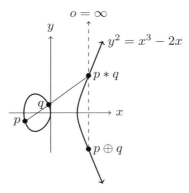

Proposition 3.29. *Let (E, o) be an elliptic curve, i.e., a non-singular projective cubic over k. Then $(E(k), \oplus)$ is an abelian group.*

Only the associativity of $\oplus$ needs checking. For a proof using nothing other than Bezout's theorem, see [4].

Definition 3.30. A *group variety* over $S = \operatorname{Spec}(k)$ is a variety $X \xrightarrow{\pi} S$ together with a section $e : S \to X$ (identity), and morphisms $\mu : X \times_S X \to X$ (group operation) and $\rho : X \to X$ (inverse) such that

(1) $\mu \circ (id \times \rho) \circ \Delta = \mu \circ (\rho \times id) \circ \Delta = e \circ \pi : X \to X$;
(2) $\mu \circ (\mu \times id) = \mu \circ (id \times \mu) : X \times_S X \times_S X \to X$;
(3) $\mu \circ (e \times id) = j_1 : S \times_S X \xrightarrow{\sim} X$, and $\mu \circ (id \times e) = j_2 : X \times_S S \xrightarrow{\sim} X$.

Clearly, for a field K extending k, $X(K)$ is a group.

Example 3.31 (Examples of algebraic groups).

(1) Additive group $\mathbb{G}_a = \mathbb{A}_k^1$. Multiplicative group $\mathbb{G}_m = \operatorname{Spec}(k[x, x^{-1}])$.
(2) $\operatorname{SL}_2(k) = \{(a, b, c, d) : ad - bc = 1\}$, with $\rho(a, b, c, d) = (d, -b, -c, a)$.
(3) $\operatorname{GL}_2(k) = \operatorname{Spec}(k[a, b, c, d, 1/(ad - bc)])$.

Proposition 3.32. *An elliptic curve (E, o) affords a group variety $(E, \oplus)$.*

In other words, the operations $\oplus : E \times E \to E$ and $\ominus : E \to E$ are morphisms.

Definition 3.33. An *abelian variety* is a connected and proper group variety (it follows that the operation is commutative, hence the name).

Thus, elliptic curves are basic examples of abelian varieties.

3.4. Riemann–Roch theorem

Definition 3.34. Let X be an integral non-singular projective curve over k. Then $\Omega_{X/k}$ is a locally free sheaf of rank 1, so pick a non-zero meromorphic section $\omega \in \Omega_{\mathbf{k}(X)/k}$. For $x \in X$, let t be the uniformiser at x, and let $f \in \mathbf{k}(X)$ be such that $\omega = f\,dt$. Define

$$v_x(\omega) = v_x(f),$$

and the resulting *canonical divisor*

$$W = \operatorname{div}(\omega) = \sum_x v_x(\omega) x.$$

The divisor W' of a different $\omega' \in \Omega_{\mathbf{k}(X)/k}$ is linearly equivalent to W, $W' \sim W$, and thus W uniquely determines a *canonical class* K_X in $\operatorname{Cl} X$.

Example 3.35. The projective line $\mathbb{P}^1$ with homogeneous coordinates $[\chi : v]$ is covered by sets $U_\chi = D(\chi)$ and $U_v = D(v)$. On U_v, we use the affine coordinate $x = \chi/v$, and on U_χ we use $y = v/\chi$. On $U_\chi \cap U_v$, we have $x = 1/y$, and $dx = -dy/y^2$. Let ω be the meromorphic section of $\Omega_{\mathbb{P}^1}$ given by $\omega = dx$ on U_v and $\omega = -dy/y^2$ on U_χ. For any point $P = [a : 1] \in U_v$, the uniformiser is $x - a$, so $\omega = dx = 1 \cdot d(x - a)$ and $v_P(\omega) = 0$. On the other hand, at $P_\infty = [1 : 0] \in U_\chi$, the uniformiser is y, so $v_{P_\infty}(\omega) = -2$. Thus, the canonical divisor is $\operatorname{div}(\omega) = -2 P_\infty$.

Example 3.36. Let $P(x) = x^3 + ax + b \in k[x]$, $\operatorname{char}(k) \neq 2, 3$. Assume that the projective curve X associated with the affine equation $y^2 = P(x)$ is regular, i.e., an elliptic curve. Thus, the *invariant differential* $\omega = dx/y = 2dy/P'(x)$ is clearly regular everywhere on the given affine piece (since P and P' cannot vanish simultaneously). By Exercise 14, ω is regular even at the point at infinity, so we conclude that the canonical divisor is $\operatorname{div}(\omega) = 0$.

Definition 3.37. Let D be a divisor on X, and write
$$L(D) = \{f \in \mathbf{k}(X) : (f) + D \geq 0\}.$$

A theorem of Riemann shows that these are finite dimensional vector spaces over k, so we can define
$$l(D) = \dim L(D).$$

Remark 3.38. Note that f and f' define the same divisor if and only if $f' = \lambda f$, for some $\lambda \neq 0$, so we have a bijection
$$\{\text{effective divisors} \sim D\} \leftrightarrow \mathbb{P}(L(D)).$$

Definition 3.39. The *genus* of a curve X is $l(K_X)$.

Theorem 3.40 (Riemann–Roch). *Let D be a divisor on a projective non-singular curve X of genus g over an algebraically closed field k. Then*
$$l(D) - l(K_X - D) = \deg(D) + 1 - g.$$

Consequently, $\deg(K_X) = 2g - 2$. From Examples 3.35 and 3.36, we see that $\mathbb{P}^1$ has genus 0, and elliptic curves are of genus 1.

4. Weil Conjectures

4.1. *The zeta function*

Definition 4.1. Let X be a variety over a finite field $k = \mathbb{F}_q$. Its *zeta function* is the formal power series
$$Z(X/\mathbb{F}_q, T) = \exp\left(\sum_{n \geq 1} \frac{|X(\mathbb{F}_{q^n})|}{n} T^n\right).$$

Example 4.2.

- Let $X = \mathbb{A}^N_{\mathbb{F}_q}$. We have $|\mathbb{A}^N_{\mathbb{F}_q}(\mathbb{F}_{q^n})| = q^{nN}$, so
$$Z(\mathbb{A}^N_{\mathbb{F}_q}, T) = \exp\left(\sum_{n \geq 1} \frac{(q^N T)^n}{n}\right) = \frac{1}{1 - q^N T}.$$

- For $X = \mathbb{P}_{\mathbb{F}_q}^N$,

$$\mathbb{P}_{\mathbb{F}_q}^N(\mathbb{F}_{q^n}) = \frac{q^{n(N+1)} - 1}{q^n - 1} = 1 + q^n + q^{2n} + \cdots + q^{Nn}, \text{ so}$$

$$Z(\mathbb{P}_{\mathbb{F}_q}^N/\mathbb{F}_q, T) = \exp\left(\sum_{n \geq 1} \frac{T^n}{n} \sum_{j=0}^N q^{nj}\right) = \prod_{j=0}^N Z(\mathbb{A}_{\mathbb{F}_q}^j, T) = \prod_{j=0}^N \frac{1}{1 - q^j T}.$$

Suppose X is over $\mathbb{F}_q$, consider the algebraic closure $\bar{\mathbb{F}}_q$ of $\mathbb{F}_q$, and the Frobenius automorphism $F_q : \bar{\mathbb{F}}_q \to \bar{\mathbb{F}}_q$, $F_q(x) = x^q$. Then F_q acts on $X(\bar{\mathbb{F}}_q) = \mathrm{Hom}(\mathrm{Spec}(\bar{\mathbb{F}}_q), X)$ by precomposing with ${}^aF_q$. Intuitively, if X is affine in $\mathbb{A}^N$, then

$$F_q(x_1, \ldots, x_N) = (x_1^q, \ldots, x_N^q).$$

Remark 4.3. The $\mathbb{F}_{q^n}$-points are fixed points of the nth power of F_q,

$$X(\mathbb{F}_{q^n}) = \mathrm{Fix}(F_q^n).$$

Remark 4.4. A closed point $x \in X$ corresponds to an F_q-orbit of an $\bar{x} \in X(\bar{\mathbb{F}}_q)$, and $[\mathbf{k}(x) : \mathbb{F}_q] = |\{\text{orbit of } \bar{x}\}| = \min\{n : \bar{x} \in X(\mathbb{F}_{q^n})\}$.

Definition 4.5. Write X^0 for the set of closed points of X. For $x \in X^0$, let

$$\deg(x) = [\mathbf{k}(x) : \mathbb{F}_q], \qquad Nx = q^{\deg(x)}.$$

Recall Riemann's definition,

$$\zeta(s) = \sum_{n \geq 1} n^{-s} = \prod_{p \in \mathrm{Specm}(\mathbb{Z})} (1 - p^{-s})^{-1}.$$

Lemma 4.6 (Euler's product formula). *The zeta function can be written*

$$Z(X/\mathbb{F}_q, T) = \prod_{x \in X^0} (1 - T^{\deg(x)})^{-1}.$$

The variable change $T \leftarrow q^{-s}$ yields a form analogous to Riemann's,

$$Z(X/\mathbb{F}_q, q^{-s}) = \prod_{x \in X^0} (1 - Nx^{-s})^{-1}.$$

The proof is an exercise upon remarking that

$$|X(\mathbb{F}_{q^n})| = \sum_{r \mid n} r \cdot |\{x \in X^0 : \deg(x) = r\}|.$$

Theorem 4.7 (Weil's conjectures). *Let X be a smooth projective variety of dimension d over $k = \mathbb{F}_q$, $Z(T) = Z(X/k, T)$. Then*

(1) Rationality. *The function $Z(T)$ is rational.*
(2) Functional equation. *The zeta function satisfies*
$$Z\left(\frac{1}{q^d T}\right) = \pm T^\chi q^{\chi/2} Z(T),$$
where χ is the 'Euler characteristic' of X.
(3) Riemann hypothesis. *We can write*
$$Z(T) = \frac{P_1(T) P_3(T) \cdots P_{2d-1}(T)}{P_0(T) P_2(T) \cdots P_{2d}(T)},$$
where each $P_i(T)$ has integral coefficients and constant term 1, and
$$P_i(T) = \prod_j (1 - \alpha_{ij} T),$$
where α_{ij} are algebraic integers with $|\alpha_{ij}| = q^{i/2}$. The degree of P_i is the 'i-th Betti number' of X.

The use of 'Euler characteristic' and 'Betti numbers' suggests that the *arithmetical* situation is controlled by the classical *geometry* of X.

Weil posed these conjectures in 1949, and proved they hold for curves in [10]. The rationality conjecture and the functional equation were first proved by Dwork using methods of p-adic analysis in [3] (which, *a posteriori*, was a rudimentary form of Monsky–Washnitzer/crystalline/rigid cohomology). A much more geometric/conceptual cohomological proof of those conjectures by Grothendieck was a culmination of the development of étale l-adic cohomology through volumes of SGA [6]. The Riemann hypothesis turned to be much more difficult, and it was proved by Deligne in [2].

4.2. *Rationality for curves*

We shall give an 'elementary' (even motivic) proof of the rationality conjecture for curves dependent only on the Riemann–Roch theorem.

Let us start by extending the theory of divisors to a non-algebraically closed base field.

Definition 4.8. Let X be a curve over k.

- $\operatorname{Div}(X)$ is the free abelian group generated by the closed points of X.

- For $D = \sum_i n_i x_i \in \text{Div}(X)$, let
$$\deg(D) = \sum_i n_i \deg(x_i).$$

- Write $\text{Div}(n) = \{D \in \text{Div}(X) : \deg(D) = n\}$ and $\text{Cl}(n) = \text{Div}(n)/\sim$.

Let X be a smooth projective curve over $\mathbb{F}_q$ now. Using Riemann–Roch theorem, if $\deg(D) > 2g - 2$, then $\deg(K_X - D) < 0$ so $l(K_X - D) = 0$ and thus
$$l(D) = \deg(D) + 1 - g.$$

Therefore, for $n > 2g - 2$, the number E_n of effective divisors of degree n is
$$\infty > E_n = \sum_{\bar{D} \in \text{Cl}(n)} \frac{q^{l(D)} - 1}{q - 1} = \sum_{\bar{D} \in \text{Cl}(n)} \frac{q^{n+1-g} - 1}{q - 1} = |\text{Cl}(n)| \frac{q^{n+1-g} - 1}{q - 1}.$$

In particular, $|\text{Cl}(n)| < \infty$.

Suppose the image of $\deg : \text{Div}(X) \to \mathbb{Z}$ is $d\mathbb{Z}$ (we will see later that $d = 1$). Choosing some $D_0 \in \text{Div}(d)$ defines an isomorphism
$$\text{Cl}(n) \xrightarrow{\sim} \text{Cl}(n + d)$$
$$D \longmapsto D_0 + D,$$

and therefore
$$|\text{Cl}(n)| = \begin{cases} J & \text{if } d|n, \\ 0 & \text{otherwise,} \end{cases}$$

where $J = |\text{Cl}(0)|$ is the number of rational points on the Jacobian of X. Note that $d | 2g - 2$ since $\deg(K_X) = 2g - 2$. Therefore,

$$Z(X/\mathbb{F}_q, T) = \prod_{x \in X^0} (1 - T^{\deg(x)})^{-1} = \sum_{D \geq 0} T^{\deg(D)} = \sum_{n \geq 0} E_n T^n$$
$$= \sum_{\substack{n=0 \\ d|n}}^{2g-2} T^n \sum_{\bar{D} \in \text{Cl}(n)} \frac{q^{l(D)} - 1}{q - 1} + \sum_{\substack{n=2g-2+d \\ d|n}}^{\infty} T^n J \frac{q^{n+1-g} - 1}{q - 1}$$
$$= Q(T) + \frac{J}{q-1} T^{2g-2+d} \left[\frac{q^{g-1+d}}{1 - (qT)^d} - \frac{1}{1 - T^d} \right],$$

so $Z(X/\mathbb{F}_q, T)$ is a rational function in T^d with first-order poles at $T = \xi$, $T = \frac{\xi}{q}$ for $\xi^d = 1$.

Lemma 4.9 (Extension of scalars). *We have*
$$Z(X \times_{\mathbb{F}_q} \mathbb{F}_{q^r}/\mathbb{F}_{q^r}, T^r) = \prod_{\xi^r=1} Z(X/\mathbb{F}_q, \xi T).$$

Proposition 4.10. *In the above calculation, if X is geometrically connected, then $d = 1$.*

Proof. By an analogous argument, $Z(X \times_{\mathbb{F}_q} \mathbb{F}_{q^d}/\mathbb{F}_{q^d}, T^d)$ has a first-order pole at $T = 1$. Using extension of scalars and the fact that $Z(X/\mathbb{F}_q, T)$ is a function of T^d, we get
$$Z(X \times_{\mathbb{F}_q} \mathbb{F}_{q^d}/\mathbb{F}_{q^d}, T^d) = \prod_{\xi^d=1} Z(X/\mathbb{F}_q, \xi T) = Z(X/\mathbb{F}_q, T)^d.$$

Comparing poles, we conclude that $d = 1$. □

Resuming the calculation with $d = 1$, we see that
$$Z(X/\mathbb{F}_q, T) = Q(T) + \frac{J}{q-1} T^{2g-1} \left[\frac{q^g}{1-qT} - \frac{1}{1-T} \right],$$

where $Q(T)$ is a polynomial of degree $2g - 2$. Reducing to the common denominator $(1 - T)(1 - qT)$, we obtain the following theorem.

Theorem 4.11 (Rationality for curves). *Let X be a smooth projective curve over $\mathbb{F}_q$. Then, for some algebraic integers α_i,*
$$Z(X/\mathbb{F}_q, T) = \frac{\prod_{i=1}^{2g}(1-\alpha_i T)}{(1-T)(1-qT)}.$$

Remark 4.12. By inspecting the above calculation of $Z(X/\mathbb{F}_q, T)$, using Riemann–Roch theorem, one can deduce the functional equation
$$Z\left(X/\mathbb{F}_q, \frac{1}{qT}\right) = q^{1-g} T^{2-2g} Z(X/\mathbb{F}_q, T).$$

4.3. Cohomological interpretation of Weil conjectures

Let X be a variety of dimension d over $k = \mathbb{F}_q$, $\bar{X} = X \times_k \bar{k}$ and let $F : \bar{X} \to \bar{X}$ be the Frobenius morphism. Fix a prime $l \neq p = \text{char}(k)$.

There exist *l*-adic *étale cohomology groups* (with compact support)
$$H^i(X) = H^i_c(\bar{X}, \mathbb{Q}_l), \quad i = 0, \ldots, 2d$$
which are finite-dimensional vector spaces over $\mathbb{Q}_l$ and F induces morphisms $F^* : H^i(X) \to H^i(X)$ and we have the *Lefschetz fixed-point formula*
$$|X(\mathbb{F}_{q^n})| = |\operatorname{Fix}(F^n)| = \sum_{i=0}^{2d}(-1)^i \operatorname{tr}(F^{*n}|H^i(X)).$$
Substituting this into the definition of the zeta function yields
$$Z(X,T) = \exp\left(\sum_{n\geq 1}\frac{T^n}{n}\sum_{i=0}^{2d}(-1)^i\operatorname{tr}(F^{*n}|H^i(X))\right)$$
$$= \prod_{i=0}^{2d}\left[\exp\left(\sum_{n\geq 1}\operatorname{tr}(F^{*n}|H^i(X))\frac{T^n}{n}\right)\right]^{(-1)^i}$$
$$= \prod_{i=0}^{2d}\left[\det(1-TF^*|H^i(X))\right]^{(-1)^{i+1}},$$
an alternating product of the characteristic polynomials of the Frobenius on cohomology, demonstrating Weil's *rationality* quite explicitly.

The *functional equation* follows from the rationality and the *Poincaré duality*; there is a functorial 'orientation' isomorphism $H^{2d}(X) \xrightarrow{\sim} \mathbb{Q}_l$ which yields a non-degenerate cup-product pairing
$$H^i(X) \times H^{2d-i}(X) \longrightarrow H^{2d} \simeq \mathbb{Q}_l.$$

The *Riemann hypothesis* was proved in a tour de force of Deligne, and it is difficult to summarise his proof in a single sentence in order to explain where the square root comes from. Intuitively, both Weil's proof for curves and Grothendieck conjectural proof using standard conjectures use the Cauchy–Schwarz inequality for a certain very abstract bilinear form.

5. Further Reading

Algebraic geometry is based on the rigour of commutative algebra, and the classic [1] will serve as a great initial source.

For a first encounter with algebraic geometry, the reader might prefer to start with [4], which focuses on curves and proves the Riemann–Roch theorem we rely on in these notes.

The most widely recommended textbook for 'serious' algebraic geometry is [7], and the present notes may serve as a brief introduction to it. We highly recommend Grothendieck's EGA [5], which is encyclopaedic, but very readable. Milne's webpage http://www.jmilne.org/ contains an abundance of freely available material, including notes on algebraic geometry.

Finally, the ever-growing [8] almost surely supersedes all of the above material, and will entertain the reader ad infinitum.

6. Exercises

1. Prove Proposition 2.3 and Proposition 2.17.
2. Fill in the details in the derivation of the Nullstellensatz from the weak Nullstellensatz in the proof of Proposition 2.7 (the 'Rabinowitsch trick').
3. If $Y \subseteq Z \subseteq \mathbb{A}^n$ are algebraic sets, show that every irreducible component of Y is contained in an irreducible component of Z.
4. Let $Y = V(x^2 - yz, xz - x) \subseteq \mathbb{A}^3$. Show that Y is a union of three irreducible components and find their prime ideals.
5. Let $I = (x^2 - y^3, y^2 - z^3) \subseteq k[x, y, z]$. Show that $V(I)$ is irreducible.
6. Verify Remark 2.30 and the claims in Example 2.33.
7. The *twisted cubic curve* $Y \subseteq \mathbb{A}^3$ over an algebraically closed field k is the set $\{(t, t^2, t^3) : t \in k\}$. Show that it is an affine variety of dimension 1 (an affine curve). Consider the projection $\pi : \mathbb{A}^3 \to \mathbb{A}^2$, $(x, y, z) \mapsto (y, z)$. Show that $Z = \pi(Y)$ is a plane algebraic curve, which is not isomorphic to Y. On the other hand, show that the restriction of π to $Y \setminus \{(0, 0, 0)\}$ is an isomorphism onto $Z \setminus \{(0, 0)\}$.
8. Let k be an algebraically closed field. Given $n, d > 0$, let $M_0, M_1, \ldots, M_N$ be all the monomials of degree d in variables $x_0, \ldots, x_n$, where $N = \binom{n+d}{n} - 1$. We define the *d-uple embedding* $\rho_d : \mathbb{P}^n \to \mathbb{P}^N$,

$$(a_0, \ldots, a_n) \mapsto (M_0(a), \ldots, M_N(a)).$$

Let $\theta : k[y_0, \ldots, y_N] \to k[x_0, \ldots, x_n]$ be the homomorphism defined by sending y_i to M_i, and let $\mathfrak{a}$ be the kernel of θ. Show that $\mathfrak{a}$ is a homogeneous prime ideal, and that ρ_d is an isomorphism onto its image $V(\mathfrak{a})$.
9. The *projective twisted cubic* is the image $\tilde{Y}$ of the 3-uple embedding $\mathbb{P}^1 \to \mathbb{P}^3$. Find the generators of the homogeneous ideal $I(\tilde{Y})$. Compare the minimal number of generators of $I(\tilde{Y})$ and $I(Y)$ from Exercise 7.

10. Analyse the proof of Proposition 2.43 found in [7]. Prove Lemma 2.51.
11. For $k = \mathbb{C}$, deduce Remark 2.82 from Liouville's theorem in complex analysis.
12. Prove Example 2.87 using the *going up* theorem of Cohen–Seidenberg: if B is an integral extension of A, then $\mathrm{Spec}(B) \to \mathrm{Spec}(A)$ is onto.
13. Prove the statement ('It is enough...') following Definition 3.2.
14. Verify that the canonical divisor in Example 3.36 is zero.
15. Prove Lemma 4.9.
16. Let X be the *Pell conic* with equation $x^2 - \delta y^2 - 4 = 0$ over a field $k = \mathbb{F}_q$ of odd characteristic, $\delta \in k^\times$. Compute the zeta function of X.
17. Follow the strategy from Remark 4.12 to prove the functional equation for the zeta function.

6.1. *Solutions and hints to selected exercises*

4. The irreducible decomposition is $Y = V(x,y) \cup V(x,z) \cup V(z-1, x^2-y)$.
7. The kernel of the epimorphism $\varphi : k[x,y,z] \to k[t]$ given by $x \mapsto t$, $y \mapsto t^2$, $z \mapsto t^3$ is the prime (since $k[t]$ is a domain) ideal $I(Y)$, clearly containing $(x^2 - y, x^3 - z)$. Thus $\overline{Y} = V(I(Y)) \subseteq V(x^2 - y, x^3 - z) = Y$, whence Y is closed and irreducible and $I(Y) = (x^2 - y, x^3 - z)$ by Nullstellensatz. Moreover, φ yields an isomorphism $\mathcal{O}_Y \to k[t]$ so $Y \simeq \mathbb{A}^1$ is a curve. We get $Z = V(y^3 - z^2)$, which is singular at $(0,0)$ so it is not isomorphic to Y.
9. The curve $\tilde{Y}$ is the image of $\rho_3 : \mathbb{P}^1 \to \mathbb{P}^3$ defined in Exercise 8 by $[s:t] \mapsto [s^3 : s^2 t : st^2 : t^3]$. The generators of its ideal in homogeneous coordinates $[x:y:z:w]$ on $\mathbb{P}^3$ are $\{xz - y^2, yw - z^2, xw - yz\}$.
14. The curve X is given by $v^2 \zeta = x^3 + a x \zeta^2 + b \zeta^3$ in homogeneous coordinates so that, using affine coordinates $x = \chi/\zeta$, $y = v/\zeta$ on $D(\zeta)$, the equation for $X \cap D(\zeta)$ becomes $y^2 = P(x)$ as in the example. The point at infinity $[0:1:0]$ lies in $D(v)$, where, using coordinates $u = \chi/v$, $v = \zeta/v$, the equation becomes $v = u^3 + auv^2 + bv^3$. On $D(\zeta) \cap D(v)$, we have $x = u/v$, $y = 1/v$, $dy = -dv/v^2$ and $v = v^3 P(u/v)$. Differentiating the equations, we see that a global differential ω can be defined by $\omega = dx/y = 2dy/P'(x)$ on $D(\zeta)$, and by $\omega = -2dv/(3u^2 + av^2) = 2du/(3bv^2 + 2auv - 1)$ on $D(v)$, and these expressions agree on $D(\zeta) \cap D(v)$.
16. If δ is a square in $k = \mathbb{F}_q$, X is isomorphic to the hyperbola $xy = 1$ so $|X(k)| = |k^\times| = q - 1$. If δ is not a square, draw lines of varying slopes α through $P = (2,0)$, to count the other points of $X(k)$ (with

$x = (\alpha^2\delta + 1)/(\alpha^2\delta - 1))$, and conclude that $|X(k)| = q + 1$. Hence, writing χ for the quadratic character on $k^\times$ (so that $\chi(\delta) = 1$ if δ is a square in k, and $\chi(\delta) = -1$ otherwise), we have the formula $|X(\mathbb{F}_{q^n})| = q^n - \chi(\delta)^n$, which yields that $Z(X/k, T) = (1 - \chi(\delta)T)/(1 - qT)$.

References

[1] M. F. Atiyah and I. G. Macdonald, *Introduction to Commutative Algebra*. Addison-Wesley Publishing Company (1969).
[2] P. Deligne, La conjecture de Weil. I. *Publ. Math. Inst. Hautes Études Sci.* **43**, 273–307 (1974).
[3] B. Dwork, On the rationality of the zeta function of an algebraic variety. *Amer. J. Math.* **82**, 631–648 (1960).
[4] W. Fulton, *Algebraic Curves*. Advanced Book Classics, Addison-Wesley Publishing Company, Redwood City, CA (1989).
[5] A. Grothendieck, Éléments de géométrie algébrique. I–IV. *Publ. Math. Inst. Hautes Études Sci.* **4, 8, 11, 17, 20, 24, 28 and 32** (1960–1967).
[6] A. Grothendieck, *Séminaire de Géométrie Algébrique du Bois Marie*. SGA1, SGA3, SGA4, SGA4.5, SGA5, SGA6, SGA7. Vols. 224, 151, 152, 153, 269, 270, 305, 569, 589, 225, 288 and 340, Lecture Notes in Mathematics, Springer-Verlag, Berlin (1960–1977).
[7] R. Hartshorne, *Algebraic Geometry*. Graduate Texts in Mathematics, No. 52. Springer-Verlag, New York (1977).
[8] The Stacks Project Authors. Stacks project. `http://stacks.math.columbia.edu` (2016).
[9] I. Tomašić, Model theory. In S. Bullett, T. Fearn and F. Smith (eds), *Algebra, Logic and Combinatorics*, LTCC Advanced Mathematics Series, Vol. 3. World Scientific (2016).
[10] A. Weil, *Sur les courbes algébriques et les variétés qui s'en déduisent*. Actualités Scientifiques et Industrielles, No. 1041. Hermann et Cie., Paris (1948).

Chapter 2

Introduction to the Modular Group and Modular Forms

W. J. Harvey

Mathematics Department
King's College London, London, UK
bill.harvey@kcl.ac.uk

The modular group $SL(2, \mathbf{Z})$ is a jewel and an engine of mathematics, a beautiful object to study in its own right but also a powerful tool widely applicable throughout the world of mathematics, particularly within number theory, topology and theoretical physics, a rich source of insights into how intricate ideas of group theory, geometry and analysis blend and interact. This short course aims to provide an elementary introduction to the group itself and to some of its activities in arithmetic, geometry and complex analysis. It is at best a superficial glimpse of the treasures which await the keen explorer.

1. Introduction to the Main Characters

1.1. *Prologue*

In this chapter, we introduce one of the most pervasive ingredients of contemporary mathematics, one which has even been included in a short list of five essential pieces in the practitioner's toolkit (addition, subtraction, multiplication, division and modular forms). As a preliminary motivation, we describe briefly the main lines of our approach, leaving formal definitions to later sections.

A *modular function* is a complex function f on the upper half-plane $\mathcal{U}$ which is invariant under the action of the modular group $SL(2, \mathbb{Z})$ as fractional linear mappings: thus,

$$f\left(\frac{az+b}{cz+d}\right) = f(z) \quad \text{for all } z \in \mathcal{U}, \text{ and } \begin{pmatrix} a & b \\ c & d \end{pmatrix} \in \mathrm{SL}(2,\mathbb{Z}).$$

The name comes from the classical problem of the *moduli* of elliptic curves, and both the group and the associated functions and forms have been in constant use in number theory since the modular j-function furnished a key ingredient in the formulation (and proof) of *complex multiplication* for elliptic curves, extending the Kronecker–Weber Theorem — which states that *all abelian extensions of the rational field are generated by roots of unity (i.e., special values of e^z)* — to imaginary quadratic number fields and special values of $j(z)$.

The same j-function provides the backdrop for one of the most celebrated recent pieces of conjectural magic, the *Moonshine conjectures* of Conway and Norton [10] connecting the coefficients of the q-expansion of j to the Monster simple group. These were finally confirmed by R. Borcherds [6] after an epic voyage through the high seas of Vertex Operator Algebras in quantum field theory and an illustration of fundamental links between the modular group and theoretical physics.

Another ingredient in this mathematical cocktail is the hyperbolic geometry of the upper half-plane, which is inherited by all *modular quotient surfaces* of the form $\mathcal{U}/\Gamma$ with Γ a finite index subgroup of $\Gamma(1)$. In the arithmetic theory of elliptic curves, the conjecture of Taniyama–Weil postulated a canonical relation between elliptic curves over $\mathbb{Q}$ and certain modular surfaces $X_0(N)$. This was famously proved by A. Wiles and R. Taylor, and with it, as a consequence, Fermat's last theorem [35]. It is only the latest in a long history of interactions between the modular group, hyperbolic geometry and arithmetic; we shall describe briefly another one, less widely known, which concerns very directly the geometry of the upper half-plane, the class of Ford circles and the theory of Diophantine approximation. Yet again, we encounter an action of the modular group which chimes with some mathematics of current relevance.

In this first section, we set out the necessary background for these developments, give some important examples of modular forms and functions and place them in the general context needed to discuss several important applications. Both group theory and complex function theory are involved, as the group $\mathrm{SL}(2,\mathbb{Z})$ itself exhibits algebraic structural properties which have inspired some exciting modern developments, as in M. Gromov's definition of hyperbolic groups and in Geometric Group Theory, the study of groups as geometric objects in their own right. For more on this topic, the reader may consult I.M. Chiswell's chapter in this volume. Due to considerations of space and time, few proofs could be included but adequate references are provided for the reader to pursue further all the topics mentioned here.

1.2. The modular group and Möbius transformations in the plane

We begin with a definition. The *modular group* is the subgroup $SL(2, \mathbb{Z})$ of the matrix group $GL(2, \mathbb{R})$ consisting of matrices with integer entries and determinant 1.

The matrices $\{\pm \mathrm{Id}\}$ commute with all elements of $SL(2, \mathbb{Z})$ and simple calculations show that no other element has this property: thus, the subgroup $\langle \pm \mathrm{Id} \rangle$ forms the *centre* ($\cong \mathbb{Z}/2$) of the group. The quotient $PSL(2, \mathbb{Z})$ is denoted $\Gamma(1)$, and is also called the (*homogeneous*) *modular group*; this slight abuse of language does not cause any problems.

The action of the Lie group $SL(2, \mathbb{R})$ on the Riemann sphere $\widehat{\mathbb{C}} = \mathbb{C} \cup \infty$ as *fractional linear* (or Möbius) transformations will dominate our study. The matrix $A = \begin{pmatrix} a & b \\ c & d \end{pmatrix}$ operates by the following rule:

$$z \mapsto T_A(z) = \frac{az+b}{cz+d}. \tag{1.1}$$

Notice that the statement 'a group G acts on a set S' means that there is a group homomorphism from G into the bijective self-maps of S, so that the group operation in G is respected by the resulting transformations of S. The reader should check that, in this case, we have

$$T_B \circ T_A(z) = T_{BA}(z), \quad \text{for all } z \in \mathbb{C}.$$

The group of real matrices preserves the subset $\widehat{\mathbb{R}} = \mathbb{R} \cup \{\infty\} \subset \widehat{\mathbb{C}}$ and, because the matrices have determinant 1, it follows that the action preserves the upper and lower half-planes separately. This can be confirmed by simple calculation as follows.

Lemma 1.1. *For any $z \in \mathbb{C}$ and $A = \begin{pmatrix} a & b \\ c & d \end{pmatrix} \in SL(2, \mathbb{R})$, we have*

$$\mathrm{Im}(T_A(z)) = \frac{\mathrm{Im}(z)}{|cz+d|^2}. \tag{1.2}$$

We concentrate attention on the upper half-plane $\mathcal{U}$ and its boundary, the completed real line $\widehat{\mathbb{R}}$, a topological circle. It is clear that the matrices

$\pm A$ have identical action on $\mathcal{U}$ and the reader can easily verify that if $B \neq \pm A$ then $T_B(z) \neq T_A(z)$ for general $z \in \mathcal{U}$. Thus, the *kernel* of this action, which is, by definition, the subgroup consisting of all elements which *act trivially* fixing all points of $\mathcal{U}$, is the centre of the group, $\langle \pm \mathbf{I} \rangle$. The quotient group $G = \mathrm{PSL}(2,\mathbb{R}) = \mathrm{SL}(2,\mathbb{R})/\langle \pm \mathbf{I}\rangle$ therefore acts faithfully on $\mathcal{U}$ and also on the boundary circle. It is shown in elementary accounts of complex analysis or hyperbolic geometry (see for example [2, 26]) that the group action on $\widehat{\mathbb{R}}$ is *triply transitive*. We record this as follows.

Theorem 1.2. *For any ordered triple of distinct points of $\widehat{\mathbb{R}}$ there is an element of $G = \mathrm{PSL}(2,\mathbb{R})$ which maps it to the (ordered) triple $0, 1, \infty$.*

1.3. Hyperbolic geometry in the upper half-plane

The *hyperbolic metric* ds_h^2 is the measure of arc length given by the formula

$$ds_h^2 = y^{-2}(dx^2 + dy^2); \tag{1.3}$$

this is a Riemannian metric on $\mathcal{U}$. It follows from Theorem 1.2 that the action of G defined in equation (1.1) is *transitive* on the points of $\mathcal{U}$. The stabiliser in G of the point i is the set of mappings T_A with

$$A = \begin{pmatrix} a & b \\ c & d \end{pmatrix} \in \mathrm{SL}(2,\mathbb{R}) \quad \text{and} \quad ai + b = i(ci + d),$$

which yields $b = -c$, $a = d$ so that A is orthogonal and hence a rotation; thus $T_A \in \mathrm{SO}(2)$. The G-action is therefore also transitive on the set of unit tangent directions at each point. We incorporate these facts as follows.

Theorem 1.3. *The upper half-plane $\mathcal{U}$ is isomorphic to the homogeneous space $\mathrm{PSL}(2,\mathbb{R})/\mathrm{SO}(2)$.*

In other words, $\mathcal{U}$ is a *Riemannian symmetric space* in the sense of differential geometry.

Invariance of the metric under the (differential of the) action by $\mathrm{SL}(2,\mathbb{R})$ is another simple calculation (see Exercises for Section 1). In fact, we have the following corollary.

Corollary 1.4. *The real Möbius transformations form the direct (i.e., sense-preserving) isometry group of $\mathcal{U}$.*

Since the action is transitive, no larger direct isometry group is possible.

<u>Notice</u> that the space $\mathcal{U}$ is Poincaré's model of *hyperbolic plane geometry*, in which geodesics between any two points are defined by circular arcs (or

line segments) orthogonal to the boundary real line. In this setting, therefore, we are able to use geometric ideas such as polygonal shape, convexity, length and area to illuminate the group activity on $\mathcal{U}$.

1.4. The upper half-plane as parameter space: Lattices in the plane

The upper half-plane serves as parameter space for a range of interesting mathematical objects. Gauss used it first to classify positive definite binary quadratic forms, and a long, illustrious list of 19th century authors established the link with complex tori and elliptic functions. We look in detail at this second aspect here, beginning with a few definitions.

First, recall that a *lattice* in a real vector space V is a subgroup Λ satisfying the following (equivalent) conditions:

(1) Λ is discrete and V/Λ is compact;
(2) Λ, discrete, generates V as an $\mathbb{R}$-space;
(3) there exists an $\mathbb{R}$-basis $\{e_1, \ldots, e_n\}$ of V which is a $\mathbb{Z}$-basis of Λ, i.e., $\Lambda = \mathbb{Z}e_1 \oplus \cdots \oplus \mathbb{Z}e_n$.

Viewing the complex plane as a $\mathbb{R}$-vector space, we see that a *lattice* in the complex plane is a discrete subgroup of the additive group of the complex numbers $\mathbb{C}$ containing two elements which are linearly independent over the real numbers.

Let $\mathcal{L}$ be the set of all lattices in $\mathbb{C}$. We denote by M the *set of all marked lattices*, which are pairs $(\lambda_1, \lambda_2) \in \mathbb{C}^*$ such that $\Im(\lambda_1/\lambda_2) > 0$, referred to as *positively oriented*. Define a map $\phi : M \to \mathcal{L}$ by the rule

$$\phi(\lambda_1, \lambda_2) = L(\lambda_1, \lambda_2) \cong \mathbb{Z}\lambda_1 \oplus \mathbb{Z}\lambda_2.$$

This map is clearly surjective.

Now any generating pair for a lattice may be changed by linear action of a matrix in $\mathrm{SL}(2, \mathbb{Z})$: let $A = \begin{pmatrix} a & b \\ c & d \end{pmatrix} \in \mathrm{SL}(2, \mathbb{Z})$ and $(\lambda_1, \lambda_2) \in M$. Write

$$\lambda_1' = a\lambda_1 + b\lambda_2, \quad \lambda_2' = c\lambda_1 + d\lambda_2.$$

Then $L(\lambda_1', \lambda_2') = L(\lambda_1, \lambda_2)$. Conversely, any two bases for a lattice can be switched by applying some matrix in $\mathrm{GL}(2, \mathbb{Z})$ and, if the two pairs are to be compatibly oriented so that both $z = \lambda_1/\lambda_2$ and $z' = \lambda_1'/\lambda_2'$ lie in $\mathcal{U}$, the matrix must have determinant 1. We have proved the result below.

Theorem 1.5. *Two elements of M determine the same lattice if and only if they are congruent modulo* $\mathrm{SL}(2,\mathbb{Z})$. *Thus,*

$$\mathcal{L} \cong M/\mathrm{SL}(2,\mathbb{Z}).$$

We now consider the *scaling action* by $\mathbb{C}^*$ on both $\mathcal{L}$ and M: for $\mu \in \mathbb{C}^*$, apply the rules $L \mapsto \mu L$ and $(\lambda_1, \lambda_2) \mapsto (\mu\lambda_1, \mu\lambda_1)$. Then generating pairs are sent to pairs $(z,1)$ with $z \in \mathcal{U}$ and we deduce the following corollary.

Corollary 1.6. *These rules induce a bijection between the space* $\mathcal{L}/\mathbb{C}^*$ *of lattice classes and* $\mathcal{U}/\mathrm{PSL}(2,\mathbb{Z})$.

By associating to a lattice L the coset space $\mathbb{C}/L$ with the quotient topology, we obtain a *complex torus* X_L. This is a compact surface of genus 1, homeomorphic to the product $S^1 \times S^1$ of two circles: a topological model for X_L is obtained by identifying opposite sides of the *fundamental parallelogram* in the plane spanned by any generating pair,

$$\Pi(\lambda_1, \lambda_2) = \{t\lambda_1 + u\lambda_2 \in \mathbb{C} \mid 0 \leq t, u \leq 1\},$$

and

$$X_L \cong \Pi(\lambda_1, \lambda_2)/\{(t,0) \sim (t,1), (0,u) \sim (1,u) \text{ for } 0 \leq t, u \leq 1\}.$$

Local complex coordinate charts for X_L are obtained by projecting small open discs from the covering space $\mathbb{C}$ to the quotient.

Notice that this produces a *complex structure* on the torus, that is, a collection of complex charts covering X_L which are related on intersections by bijective holomorphic transition maps in the complex plane — in this case, complex affine maps between covering discs — making it a compact complex manifold, a *Riemann surface*. Then, by the methods of complex geometry, as we describe below, one can produce an explicit equation for an algebraic curve representing X_L, which is called an *elliptic curve*, because such objects first emerged from the study of 'elliptic integrals' by Euler, following earlier work of Fagnano. For this and more on the historical side of complex function theory that underpins this chapter, the reader may consult the excellent text by Siegel [32].

We consider next the question of describing in a suitably invariant manner all possible shapes of complex tori, surfaces of genus 1 with a complex analytic structure. In order to link it to key aspects of the modular group action given above, we sketch the main idea and then return to it in Section 3 for the construction of modular forms, since the classical

Weierstrass theory of elliptic (i.e., doubly periodic meromorphic) functions depends on this same standard model for a torus coming from a choice of generating set for the lattice of periods. For any point z in the upper half-plane, we obtain a lattice $L(z,1)$ denoted L_z with positively oriented generating pair $z, 1$. For instance, if we use the *square lattice* of points $\{m + ni : m, n \in \mathbb{Z}\}$, given by the set of points with integer coordinates in the plane, the lattice is the subgroup

$$L_i = \mathbb{Z} + \mathbb{Z} \cdot i \cong \pi_1(X). \tag{1.4}$$

Underlying this, there is the important concept of a homotopy-marking for the surface, which lies at the heart of the *theory of surface deformations*, the study of varying shapes of torus, which can be extended to any other type of surface: a *homotopy-marking* is a homotopy class of homeomorphisms from a fixed base torus X_0 to the varying one. Thus, using the base surface $X_0 = \mathbb{C}/L_i$ obtained as above from the square lattice, a mapping to an arbitrary (i.e., variable) one can be given by specifying a real linear mapping f_{λ_1, λ_2} of the plane sending i to λ_1 and 1 to λ_2. Intuitively, this amounts to the effect of changing the shape of a fundamental parallelogram tile for this lattice group of plane translations. We summarise this as follows.

Theorem 1.7. *The space of shapes of complex tori is the quotient space $\mathcal{U}/\Gamma(1)$.*

The method delivers a precise description of the space $\mathcal{U}$ of all geometric shapes for a torus, the first *space of moduli*, a ground-breaking step in algebraic geometry and the precursor of a widespread pattern of organised structural designs for Riemann surfaces in general known as *Teichmüller Theory* and with modified applications to other types of algebraic variety of certain specified kinds. This turns out to be a powerful tool with pervasive influence throughout mathematics and in theoretical science generally.

1.5. *Geometry of real Möbius transformations*

We classify the elements of $G = \mathrm{PSL}(2, \mathbb{R})$ into types, in a manner which respects conjugacy classes, and study the corresponding mappings of $\mathcal{U}$. All these types occur in the modular group $\Gamma(1)$.

For this division into distinct geometric types of mapping, one can either use the linear algebraic classification of real (invertible) matrices by eigenvalues or go by the (closely related) fixed point properties, which motivate the present approach. Recall first that any real or complex Möbius transformation distinct from the identity has either one or two fixed points in the Riemann sphere $\mathbb{CP}^1$; the matrix entries force restrictions in the geometric types which occur.

Definition 1.8. A real Möbius map or isometry of $\mathcal{U}$ is called *elliptic* if it has one fixed point inside $\mathcal{U}$. It is *parabolic* if it has one boundary fixed point, *hyperbolic* if it fixes two boundary points.

Real elliptic elements have *two complex conjugate fixed points*, one in each of the upper and lower half-planes.

Typical examples of *parabolic* transformations are the real translations $T(z) = z + b$, with $b \neq 0$. Each translation fixes the point ∞ and preserves (as a set) every horizontal line in $\mathcal{U}$: this is the family of *horocycles* at ∞. Any parabolic map is conjugate to a translation and leaves invariant the family of circles tangent to the boundary circle at the fixed point: these facts follow easily from the geometric properties of Möbius transformations.

Hyperbolic transformations are conjugate to real *dilations* $V_\lambda(z) = \lambda z$, with $\lambda > 1$. Each one fixes a pair of boundary points, and preserves the hyperbolic geodesic line joining them, acting on this line, the *axis* of the hyperbolic map, as a (hyperbolic) translation from the repelling fixed point towards the attracting fixed point (in the above example V_λ fixes the origin 0 and ∞ and the axis is the vertical imaginary one).

Note. The *fixed points* ξ of a Möbius map T_A with $A = \begin{pmatrix} a & b \\ c & d \end{pmatrix}$ satisfy

$$c\xi^2 + (d - a)\xi - b = 0.$$

Thus, the classification into types turns on the value of the *discriminant* of A,

$$(d - a)^2 + 4bc = (\text{Trace}^2 - 4\det)(A);$$

if $A \in \text{PSL}(2, \mathbb{R})$, then we have

- T_A is elliptic if and only if $\text{Trace}^2 A < 4$;
- T_A is parabolic if and only if $\text{Trace}^2 A = 4$;
- T_A is hyperbolic if and only if $\text{Trace}^2 A > 4$.

We can see easily that the modular group $\Gamma(1)$ contains each kind of mapping. In particular, the involution $U(z) = -1/z$ is elliptic and translation $T(z) = z + 1$ is parabolic; these two will play a prominent part in our discussion. The composition $T \circ U = S$ is also elliptic, fixing

$\rho = e^{i\pi/3}$. For hyperbolic elements, which predominate, in the sense that there are infinitely many conjugacy classes of them in $\Gamma(1)$, there is a powerful geometric ingredient in their contribution, arising from the dynamical behaviour they determine both in the hyperbolic plane and in $\mathbb{R}^2$; see for instance the following note.

Diversion: Anosov mappings of the torus. An interesting phenomenon relating to Ergodic Theory occurs when the linear map of the plane given by a hyperbolic matrix $A \in \mathrm{SL}(2,\mathbb{Z})$ is *iterated* and projected to the quotient (square) torus X_0: if we apply successive higher powers of the matrix A^n, the ensuing distortion of the plane becomes increasingly chaotic when viewed on the quotient torus. This is illustrated by the so-called *Arno'ld's cat mapping*, given by the matrix

$$A = \begin{bmatrix} 1 & 1 \\ 1 & 2 \end{bmatrix}.$$

This corresponds to a Möbius group element T_A of hyperbolic type with fixed points at $(1 \pm \sqrt{5})/2 \in \mathbb{R}$. The matrix determines a linear transformation of the plane, preserving the square grid lattice, and generates a semigroup of positive iterates, $A \circ A = A^2, A \circ A^2 = A^3, \ldots$, which distorts the unit square F (thought of as a tile within the plane) by stretching in one eigendirection and shrinking in the other transverse one, to give a sequence of image parallelograms $A^m(F)$ with corners at lattice points. Thus, $A(F)$ has corners at O, $A(\mathbf{e_1})$, $A(\mathbf{e_2})$ and $A(\mathbf{e_1} + \mathbf{e_2})$. The image has area 1 and projects surjectively onto the quotient torus, by reduction mod 1 in each coordinate: this effected by cutting up the first image parallelogram $A(F)$ where it overlaps the lattice of square tiles and then translating these pieces back to congruent parts of the original F. This mapping $\widetilde{A}$ divides F into four triangular pieces along the main diagonal and two parallel line segments, each with slope 2, through the origin and the opposite diagonal corner; this extends (if desired) to a map of the plane $\widetilde{A}$ by periodicity.

The same linear mapping A is now applied again (iterated) and projected, generating a more and more stretched out (and so more fragmented) tiling dissection pattern as m grows. However, when one places on the original tile a photographic image, which is produced by an array of $n \times n$ black or white pixelated dots, the effect of each map $\widetilde{A}^{\circ m} = A \circ A \circ \ldots A$ is seen as just a permutation of the dots, and thus some (possibly very large) power m induces the identity mapping on the finite subset of pixelated 'periodic points', although the iterated map itself is of course greatly different from the identity overall, so that the photograph appears magically unaltered, a highly paradoxical effect when viewed among the surrounding chaotic patterns — an account may be found on the Web, with illustrations, by consulting Wikipedia on 'Arnold's Cat Map'.

We note that this discussion concerns the linear action of the transformations A^m in the plane projected to the torus and has little to do with the behaviour of the corresponding hyperbolic plane isometry T_A, which operates as a translation along its axis in $\mathcal{U}$, with the fixed points of T_A, the eigenvalues of A, as the extremities of the axis. Further dynamical properties of these *Anosov maps* of the torus are discussed in V.I. Arnold's book Mathematical Methods in Classical Mechanics (Springer-Verlag, 1989) and in [4].

1.6.   *Exercises for Section 1*

(1) Show that the action of $G = \mathrm{SL}(2,\mathbb{R})$ by fractional linear maps on the upper half-plane is transitive, i.e., given any two points $z \neq w$, there is an element $\gamma \in G$ with $w = \gamma(z)$. [*Hint*: let $z = i$ first; then compose.]

Show also that the action is transitive on the tangent bundle, that is the union of all the tangent spaces at all points of $\mathcal{U}$: for any two unit tangent vectors u, v based at z, w, respectively, there is a $\gamma \in G$ with $\gamma(z) = w$ and $\gamma'(z).u = v$. [*Hint*: do Q3 first.]

(2) Show that the hyperbolic line element ds_h^2 is G-invariant.

(3) Show that the set of matrices $\gamma \in G$ which fix the point i is the compact subgroup $K = \mathrm{SO}(2,\mathbb{R}) \cong S^1$ of rotation matrices. Deduce that the upper half-plane $\mathcal{U}$ is isomorphic to the (right) K-coset space $K \backslash G$.

(4) Find all the compact subgroups of G. [*Hint*: show that this amounts to finding all the compact subgroups of the rotation subgroup K and its conjugates in G.]

(5) Show that the modular group $\mathrm{SL}(2,\mathbb{Z})$ acts transitively on the set $\mathbb{Q} \cup \infty$. Hence find parabolic elements of the modular group which fix a given rational point p/q.

Here are two exercises which tie in this group action with the general study of Lie transformation groups in differential geometry.

(6) Find all compact subgroups of the Lie group $\mathrm{SL}(2,R)$.

[*Comment.* The stabiliser of any point is a compact subgroup, conjugate to the subgroup $\mathrm{PSO}(2)$ stabilising i. If we let K be any compact subgroup of G, the set of images of i under all $T \in K$ is a compact subset C of $\mathcal{U}$ (why is this?). Conversely, given any compact set C in the upper half-plane, the set of elements $\gamma \in G$ with $\gamma(i) \in C$ is compact. But is it a subgroup? (Almost never!) So how can we pin down the compact subgroups?]

(7) Show that the action of $\mathrm{SL}(2,R)$ on $\mathcal{U}$ is *proper*: that is, prove that if K is any compact subset of $\mathcal{U}$, then the set of all g in $\mathrm{SL}(2,R)$ such that gK intersects K non-trivially is compact.
[This is the fundamental property which distinguishes actions of a Lie group on a space with compact stabilisers.]

2. Discontinuity of the Group Action: The Modular Surface

At the heart of any study of a discrete group action lies a key ingredient for coming to grips with it geometrically. This has particular force for the modular group in the present case.

2.1. *Fundamental sets for discrete groups*

Definition 2.1. Let Γ be a discrete group acting by isometries on a metric space X. A *fundamental set* for Γ is defined to be a closed set F with two key properties.

(i) the interior F^0 of F has empty intersection with each translate:

$$F^0 \cap \gamma F^0 = \emptyset \quad \text{when } \gamma \in \Gamma \setminus Id.$$

(ii) the union of all translates γF for $\gamma \in \Gamma$ covers the space X.

Thus, a fundamental set provides a tiling of the space without overlaps that is compatible with the group action. For the modular group, there is a popular and classic choice of fundamental set which we construct below.

Notice that a group which acts properly and isometrically on a metric space with discrete orbits can be provided with a fundamental set with a geometric flavour, called a *Dirichlet fundamental set*. This is the set of points F_{z_0} which are closer to a designated base point z_0 than to any other point of the Γ-orbit of z_0. Such a set is clearly convex in the sense of hyperbolic geometry, bordered by hyperbolic geodesic edges. See Exercises for Section 2 for more details.

The crucial fact about a fundamental set F in $\mathcal{U}$ for a discrete subgroup Γ such as $\Gamma(1)$ is the elementary result below, which follows immediately from the definition.

Lemma 2.2. *Let F be a fundamental set for Γ. For any point $z \in \mathcal{U}$, the Γ-orbit $\Gamma.z = \{\gamma(z) | \gamma \in \Gamma\}$ has at least one point in F and at most one point in F^0.*

Then, if the set F is of simple shape, constructive methods of topology and algebra can be applied.

In the case at hand, we use a direct approach, following [28]. First of all we mention three elements of Γ which feature in the presentation: they are

$$T(z) = z + 1; \quad U(z) = \frac{-1}{z}; \quad S = T \circ U, \ S(z) = \frac{z-1}{z}.$$

Next we concentrate attention on a certain *hyperbolic triangle*

$$\mathcal{D} = \{z \in \mathcal{U} \mid |z| > 1 \text{ and } -1/2 \leq \Re(z) \leq 1/2\}$$

with two vertices on the unit circle at the points $\rho = e^{i\pi/3}$ and $\rho^2(=\rho-1)$ and a third vertex at ∞, often called an *ideal vertex*. The edges joining these points are hyperbolic geodesic line segments: the edges to ∞ are vertical half-lines. See Fig. 1.

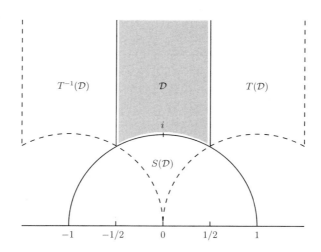

Fig. 1. A fundamental domain $\mathcal{D}$ for the modular group.

It is easily seen that U and S are *torsion elements*, with orders 2 and 3 respectively, and their fixed points are i and ρ respectively. The product $T = SU$ is the parabolic element $z \mapsto z+1$ fixing ∞.

The maps U and T determine *side-pairing transformations* in $\Gamma(1)$, taking the triangle $\mathcal{D}$ onto a neighbouring triangle which shares an edge with $\mathcal{D}$; these two maps thus enjoy properties crucial to spreading the whole action of $\Gamma(1)$ on $\mathcal{U}$. In particular, we can express all of the transforms of $\mathcal{D}$ by elements of $\Gamma(1)$ in terms of compositions given by words $W(U,T)$ in these two letters, such as

$$UT, T^2UT^{-1}, T^{-1}UUT^5U, \ldots$$

as we prove in the next section. Sometimes, as for the third word above, there are cancellations to be made, using one of two standard rules: $U^2 = \text{Id}$ and $T^{-1}T = TT^{-1} = \text{Id}$.

2.2. A fundamental domain for $\mathbf{SL}(2,\mathbb{Z})$

In fact we will prove that the subgroup Γ' of $\Gamma(1)$ generated by T and U, which of course contains $S = T \circ U$, has $\mathcal{D}$ as fundamental domain.

Theorem 2.3. (a) *For each $z \in \mathcal{U}$, there exists $\gamma \in \Gamma' = \langle T, U \rangle$ with $\gamma(z) \in \mathcal{D}$.*
(b) *If $z, z' \in \mathcal{D}$ with $z' = \gamma(z)$ and $\gamma \in \Gamma'$, then $z \in \partial\mathcal{D}$.*
(c) *For all $z \in \mathcal{D}$, the stability subgroup $\text{Stab}_\Gamma(z)$ is trivial except for the points ρ, i and $\rho - 1$.*

Proof. (a) From Lemma 1.1, we have $\Im\gamma(z) = \Im(z)/|cz+d|^2$.

Fix $z \in \mathcal{U}$. The set of all values $\{cz + d : c, d \in \mathbb{Z}\}$ is discrete. Hence the set of all values of $\Im\gamma(z)$ for $\gamma \in \Gamma'$ has a maximum at some point $z_0 \in \Gamma'.z$: this follows from the fact that there are only a finite number of pairs $c, d \in \mathbb{Z}$ with $|cz+d|$ less than a given bound.

Furthermore, any point $z \in \mathcal{U}$ has a translate $T^k(z) = z + k$ whose real part x lies in the interval $|x| \leq 1/2$. We claim that there is an element γ in Γ' with $z_1 = \gamma(z_0) \in \mathcal{D}$: for if not, then $|z_1| < 1$ and $\Im(-1/z_1) > \Im z_1 = \Im z_0$ contradicting our choice of z_0. This proves (a).

(b) Let $z, \gamma(z) \in \mathcal{D}$, with $\gamma = \begin{pmatrix} a & b \\ c & d \end{pmatrix}$. Without loss of generality, we may assume that $\Im\gamma(z) \geq \Im(z)$. Therefore $|cz+d| \leq 1$ from the earlier calculation; hence it follows that $|c| \leq 1$, since $|z| \geq 1$ and c, d are integers so $|c| \geq 2$ is impossible. Now examining in turn the three cases $c = 0$ and

$c = \pm 1$ completes the argument: for instance, if $c = 0$ then $d = \pm 1$ so γ is a translation and z lies on a vertical edge.

(c) This property follows from the case analysis in (b). □

Corollary 2.4. $\Gamma(1) = \Gamma'$: *in other words, the modular group is generated by the elements S and T. Equivalently, it is generated by the pair T and U, or by the two torsion elements S and U.*

Corollary 2.5. *The projection mapping $\pi : \mathcal{U} \to X(1) = \mathcal{U}/\Gamma(1)$ is surjective when restricted to $\mathcal{D}$: the map $\pi : \mathcal{D} \to X(1)$ is given by gluings, $\sim_T$ of the vertical edges with the map T and $\sim_U$ of the edge(s) on $|z| = 1$ with U.*

2.3. The modular surface

We consider now the topology of the quotient surface $X(1) \cong \mathcal{D}/\{\sim_T, \sim_U\}$: identifying the vertical edges via T gives an infinite cylinder, with lower end the image of a sector of the unit circle which is folded shut using the involution U, to give a topological surface homeomorphic to the plane. [Exercise: Justify this statement.]

Next we compactify this surface by adding a single extra *point at* ∞. To complete the process topologically, we need to introduce a family of open disc neighbourhoods (nbds) centred at this point.

Definition 2.6. Let $D(\infty, y_0)$ denote the subset $\{z \in \mathcal{U} : \Im(z) > y_0\}$. This is called a *horocyclic disc* (or horodisc) at ∞ for the group $\Gamma(1)$.

This set is clearly invariant under the full translation subgroup of $\mathrm{PSL}(2,\mathbb{R})$, and its intersection with the fundamental region $\mathcal{D}$ gives a fundamental region for the (discrete) translation subgroup $\mathrm{Stab}_{\Gamma(1)}(\infty) = \langle T \rangle$.

Proposition 2.7. *For each $y > 0$, the horodisc $D_y = D(\infty, y)$ is preserved setwise by the translation subgroup $\langle T \rangle$ of the modular group which acts discontinuously on it, producing as quotient a punctured disc.*

Proof. It follows from the preceding construction of fundamental region that no other translate of D_y can intersect it. For the rest, one checks that the mapping

$$z \mapsto q = e^{2\pi i z}$$

induces a holomorphic bijection from $D_y/\langle T \rangle$ to the punctured q-disc $D^*(0, r)$ where $r = e^{-2\pi y}$. □

Remarks. (1) A closer analysis of this combinatorial structure of the $\Gamma(1)$ action on $\mathcal{U}$ shows that the group has presentation

$$\langle U, S \mid U^2 = S^3 = \mathrm{Id}\rangle.$$

Thus it is isomorphic to the *free product* of the two cyclic groups C_2 and C_3. See Exercises in Section 4 for more on this aspect of combinatorial group theory.

(2) Since the modular group is generated by torsion elements U and S, it follows by a theorem of Armstrong [1] on branched coverings that the quotient surface $\mathcal{U}/\Gamma(1)$ is simply connected. Taken with the preceding theorem, which shows that the boundary of $X(1)$ is a single point, this implies that $X(1)$ is homeomorphic to the plane $\mathbb{C}$.

(3) There are two points of $X(1)$ with non-trivial stability group, the images $P(i) = \Gamma(1).i$ and $P(\rho)(= P(\rho - 1)) = \Gamma(1).\rho$ for the corners of $\mathcal{D}$. At these points, the local topology and complex structure is ramified and needs separate treatment, involving the notion of *cone point* to be discussed in the next subsection.

2.4. Cone points and cusps

Consider the projection map from $\mathcal{U}$ to the modular surface $X(1)$ in a small disc centred at i. Since the elliptic involution $U : z \mapsto -1/z$ fixes i, projection has the effect of a half-turn rotation on the tangent space at i and we see that for small $\delta > 0$, the open sets

$$N(i, \delta) = \{z : |(z - i)/(z + i)| < \delta\}$$

in the hyperbolic plane are U-invariant and project 2:1 onto disc nbds of $P(i)$ in the modular surface. Similarly, for the point ρ fixed by S of order 3, there are hyperbolic disc nbds $N(\rho, \delta)$ stabilised by the cyclic subgroup $\langle S\rangle$ and projecting 3:1 to a disc nbd of $P(\rho)$ in $X(1)$. These nbds determine the local complex analytic structure at the two special points, which are called *cone points* of the modular surface because the total angle there is less than 2π. (They are also called *ramification or branch points* of the projection from $\mathcal{U}$.)

The resulting compact surface is given a global complex analytic structure extending that of $X(1)$. We add a point (denoted P_∞) which corresponds to the orbit $\Gamma(1)(\infty) = \mathbb{Q} \cup \infty$ and need only to describe the nbds of P_∞. First, recall that the complex plane $\mathbb{C}$ has a one-point compactification $\widehat{\mathbb{C}} = \mathbb{C} \cup \infty$ known as the *Riemann sphere*; this is a surface homeomorphic to the unit sphere S^2 in $\mathbb{R}^3$. There is a standard way to set up a complex structure on S^2, using as charts the two stereographic projections of S^2 to the horizontal tangent planes at south pole S and north pole N respectively. In real coordinates $(x, y, t) \in \mathbb{R}^3$ with

$$x^2 + y^2 + t^2 = 1,$$

we embed $U_0 = S^2 - N$ and $U_1 = S^2 - S$ by the mappings

$$\phi_0(x, y, t) = \frac{x + iy}{1 - t} \quad \text{and} \quad \phi_1(x, y, t) = \frac{x - iy}{1 + t}.$$

A small calculation shows that the transition function between $\phi_0(U_0)$ and $\phi_1(U_1)$ on the intersection $U_0 \cap U_1$ is

$$\phi_1 \circ \phi_0^{-1}(z) = 1/z, \quad \text{for } z \in \mathbb{C}^*.$$

Thus, to extend the complex structure on the *modular surface* $X(1)$ we proceed as follows: add the single extra point at ∞ to obtain a closed surface $\widehat{X(1)}$, homeomorphic to the Riemann sphere. It follows then from the uniformisation theorem that this surface is analytically equivalent to the Riemann sphere.

Neighbourhoods of ∞ are the horodiscs defined above, with ∞ filling in the missing origin in the punctured discs $D(\infty, y)$: thus, if $f : X(1) \to \mathbb{C}$ is given as a function from $\mathcal{U}$ which is invariant under the translation $T : z \mapsto z + 1$, then f may be rewritten as a Fourier series $\tilde{f}(q)$ with the parameter $q = e^{2\pi i z}$ defined in some nbd $N^*(0, r) = \{0 < |q| < r = e^{-2\pi y}\}$, isomorphic to $D(\infty, y)$. The function f is then called *holomorphic at* ∞ if the function $\tilde{f}(q)$ has a removable singularity at $q = 0$.

More generally, we define the *Laurent expansion* at ∞ of a modular function $f : \mathcal{U} \to \mathbb{C}$ by viewing it as a function $\tilde{f}(q)$ which is holomorphic in some N^* and has a Laurent expansion

$$\tilde{f}(q) = \sum_{n=-\infty}^{\infty} c_n q^n.$$

The function f is called *meromorphic at* ∞ if $c_n = 0$ for all but finitely many $n < 0$.

In accord with most authors, we shall include a condition 'holomorphic at ∞' in the definition of modular form. In view of Corollary 2.4, we rephrase the definition of modular form as follows.

Definition 2.8. A *modular form of weight k for the full modular group* $\Gamma(1)$ is a function $f(z)$ with $f(z+1) = f(z)$ satisfying $f(-1/z) = z^{2k} f(z)$ with Fourier expansion

$$\tilde{f}(q) = \sum_{n=0}^{\infty} c_n q^n.$$

If $c_0 = 0$, the modular form $f(z)$ is called a *cusp form*.

2.5. Automorphic forms: The definition and examples

This is a version of the concept of modular form valid for any Fuchsian group.

Definition 2.9. An *automorphic form* of weight $k/2$ for a discrete group $\Gamma \subset \mathrm{SL}(2, \mathbb{R})$ is a function on the upper half-plane which satisfies the following functional equation with respect to the action of the group Γ.

$$f(\gamma(z)) = (cz+d)^{2k} f(z), \quad \text{for all } \gamma(z) = \frac{az+b}{cz+d} \in \Gamma. \quad (2.1)$$

Various adjectives (such as holomorphic, meromorphic, real analytic, ...) may be attached to this concept. Note that the multiplier which occurs here is linked to $(cz+d)^{-2} = (d/dz)\gamma(z)$, implying that *there is an interpretation of the formula in the case $k=1$ as a differential form of weight 1 on the quotient Riemann surface $X = \mathcal{U}/\Gamma$*, that is, a section of the cotangent bundle of X. For if we had a non-constant function F satisfying the equation (2.1) with $k = 0$, implying that $F(\gamma(z)) = F(z)$ for all $\gamma \in \Gamma$, that would project to give a genuine function on X, which implies that its derivative F' would satisfy (2.1) with $k = 1$.

We shall say that a (meromorphic) function on X satisfying the formula (2.1) defines an *automorphic form of weight k* for Γ, provided it obeys a technical condition on the growth of $F(z)$ as $z \to \infty$ which ensures that any singularity at ∞ is at worst a pole.

Some examples of modular forms (Eisenstein series) will arise naturally as functions on the space of lattices in our discussion of elliptic functions. The simplest expression we can give at present is the following convergent infinite product in the unit disc $\{|q| < 1\}$ where $q = e^{2\pi i z}$:

$$\Delta(q) = q \prod_{n=0}^{\infty} (1 - q^n)^{24} = q - 24q^2 + 252q^3 - 1472q^4 + \cdots, \quad (2.2)$$

which turns out to be a cusp form of weight 12; this function will be discussed in the next section. One can show directly using Corollary 2.4 that it satisfies (2.1): it would be difficult to carry out at this point, but not impossible (see [28, Section 4.4]).

2.6. Exercises for Section 2

(1) Show that any automorphism (i.e., biholomorphic self-mapping) of the complex plane is a complex affine map

$$T(z) = az + b, \quad a \neq 0.$$

[*Hint*: such a mapping has the property that $f(z) = 1/T(z)$ has a removable singularity at the isolated point ∞.]

If in addition T commutes with the complex conjugation operator $J : z \mapsto \bar{z}$, what can be said about T?

(2) Let T be a hyperbolic element of $\mathrm{PSL}(2,\mathbb{R})$ fixing two points a and b in $\partial\mathcal{U}$, and let ℓ be the h-axis joining a and b. If $P \in \ell$ show that the hyperbolic distance $d_h(P, T(P))$ is a constant $\lambda(T) > 0$: it is called the *translation length* of T.

How is $\lambda(T)$ related to the matrix of T?

(3) Look up the definition of *Dirichlet fundamental domain* in either [19] or [5]. Verify that for $\Gamma(1)$ with centre $z_0 = it$ on the imaginary axis and $t > 1$ the Dirichlet domain is the triangular region $\mathcal{D}$ defined above.

(4) Let $N > 1$ be a positive integer. The *principal congruence subgroup* $\Gamma(N)$ *with level* N is defined to be the kernel of the mapping '*reduction mod N*' from $\mathrm{SL}(2,\mathbb{Z})$ to $\mathrm{SL}(2,\mathbb{Z}/N\mathbb{Z})$. Show that this is an epimorphism. Show that the kernel has no torsion elements if $N > 2$.

If $N = 2$, show that the subgroup $\Gamma(2)$ has index 12. Can you calculate the index for given N prime? for general N? (See [19, Chapter 6]).

(5) (a) Write $\mathbb{C}^* = \mathbb{C} - \{0\}$ and let $\lambda \in \mathbb{C}^*$; let Λ denote the cyclic subgroup $\langle \lambda \rangle = \{\lambda^n \mid n \in \mathbb{Z}\}$. Show that the coset space $\mathbb{C}^*/\Lambda = X$ is a compact Riemann surface. Can you identify it as a topological surface? What is its universal covering space?

(b) Define a function f by the rule

$$f(z) = z \sum_{n=-\infty}^{\infty} (\lambda^{n/2} z - \lambda^{-n/2})^{-2}.$$

Prove that this sum converges uniformly on compact subsets of $\mathbb{C}^* - \Lambda$ and that the function satisfies $f(\lambda z) = f(z)$ for all z not in Λ. Discuss

the nature of the singularity of f at $z = 1$. Show that f determines a meromorphic function by projection on the Riemann surface X.
(6) Let $Q(x,y) = ax^2 + bxy + cy^2$ be a positive definite quadratic form with real coefficients. Show that it can be expressed in the form

$$Q(x,y) = |xz_1 + yz_2|^2,$$

for suitable $z_j \in \mathbb{C}$. Writing $w = z_1/z_2$, use Theorem 2.3 to deduce that there exists a form $Q' = a'x^2 + b'xy + c'y^2$ which is equivalent to Q under the linear action of $\mathrm{SL}(2,\mathbb{Z})$ on $\mathbb{C}^2$ and which satisfies the inequalities

$$a' \geq c' \geq |b'|.$$

[*Note*: This appeared first in Gauss's 'Disquisitiones Arithmeticae' (published 1801).]

(7) Prove that the formula

$$T(z) = e^{i\theta}\left(\frac{z-\alpha}{1-\overline{\alpha}z}\right)$$

defines an automorphism of the unit disc $\mathbb{D} = \{|z| < 1\}$ for all values of $\theta \in \mathbb{R}$ and $\alpha \in \mathbb{D}$.

[*Note*. Many issues in hyperbolic geometry are more conveniently viewed in the unit disc model, with the group $\mathrm{SU}(1,1)$ of all such mappings as automorphism group.]

3. Modular Forms: Eisenstein Series and Moduli of Lattices

We concentrate on forms for the full modular group $\Gamma(1)$, constructing the standard Eisenstein series $E_k(z)$ and the famous j-function. At the same time, we indicate how the methods extend to modular forms associated with any finite index subgroup Γ.

3.1. *Cusps and horodiscs: Cusp neighbourhoods*

We explain the precise complex structures to be deployed at boundary points in order to define and extend canonically the Riemann surface structure of a discrete quotient $\mathcal{U}/\Gamma$, with $\Gamma \subset \Gamma(1)$ any finite index subgroup.

Firstly, notice that the number of cusp orbits for the subgroup Γ is finite because the index $[\Gamma(1):\Gamma]$ is finite. Furthermore, each cusp P of Γ is an image $a/c = \gamma(\infty)$ of the standard cusp of $\Gamma(1)$, for some $\gamma \in \Gamma(1)$ with $\gamma = \begin{pmatrix} a & b \\ c & d \end{pmatrix}$ not in Γ. This leads to a transformation rule for setting up local coordinates near a/c, in analogy with q-expansions. The stability group of the Γ-cusp is generated by some power $\tilde{T}^k$ of the conjugate generator $\tilde{T} = \gamma T \gamma^{-1}$. This implies that the point a/c lies in the boundary of k copies of the fundamental triangle $\mathcal{D}$. Without loss of generality, we may employ a *channel of width* k as cusp nbd of P and a local coordinate at the cusp will be the corresponding transform by γ of $q' = e^{2k\pi i z}$ at ∞.

This process results in a compact Riemann surface $\widehat{X(\Gamma)}$ with a covering map π induced by the inclusion of a Γ-orbit $\Gamma.z$ in the corresponding $\Gamma(1)$-orbit: we have $\pi : \widehat{X(\Gamma)} \to \widehat{X(1)}$, the Riemann sphere, with ramification at any cusp P where $k \neq 1$.

We summarise our discussion in the following way:

Proposition 3.1. *For each finite index subgroup Γ of $\Gamma(1)$, the corresponding quotient surface $X(\Gamma) = \mathcal{U}/\Gamma$ has a natural compactification*

$$\widehat{X(\Gamma)} = X(\Gamma) \cup \{\text{cusps of } \Gamma\},$$

which defines a covering map $\pi : \widehat{X(\Gamma)} \to \widehat{\mathbb{C}}$, with branching over at most three ramification points.

The *ramification points* of $X(\Gamma)$ are at points lying over the $\Gamma(1)$-orbits of i, ρ and ∞, since these are the only places of $\mathcal{U} \cup \widehat{\mathbb{R}}$ where projection may not be locally a homeomorphism. At a ramification point of degree k, we have a projection which is of the form $z \mapsto z^k$ in local coordinates, which is k to 1 away from the centre.

Note that an automorphic form for the subgroup $\Gamma < \Gamma(1)$ which is meromorphic at all the cusps of Γ is still called a modular form; such forms will have q-expansions at a cusp of degree k which are series in powers of q^k.

3.2. Eisenstein series

These are functions $G_n(\Lambda)$ defined on the space of lattices $\mathcal{L}$ which have self-evident invariance properties, and which occur naturally, for instance, in the expansion of the Weierstrass elliptic functions.

Definition 3.2. The *Eisenstein series* is the infinite series given below:

$$G_s(\Lambda) = \sum_{\lambda \in \Lambda^*} \lambda^{-s}. \tag{3.1}$$

The summation is over $\lambda \in \Lambda - \{0\}$.

Theorem 3.3. *The series $G_s(\Lambda)$ converges absolutely for all real $s > 2$.*

Proof. This follows by the two-dimensional integral test, comparing the series with the double integral

$$\iint \frac{dx\, dy}{(x^2 + y^2)^s}$$

taken over the real plane minus a disc centred at 0, or using comparison with the standard series $\sum_{n=1}^{\infty} n^{-s+1}$. To sketch this method, we choose two generators λ_1, λ_2 of the lattice, and sum by bracketing terms together in *parallelogram layers*, which involve combinations $n_1\lambda_1 + n_2\lambda_2$ with integers n_1, n_2 such that $|n_1| + |n_2| = n$, so that in the tessellation of the plane by the parallelogram $P = P(\lambda_1, \lambda_2)$ with corners at $0, \lambda_1, \lambda_2$ and $\lambda_1 + \lambda_2$, 8 lattice points lie on a parallelogram P_1 at layer 1 from the origin 0 in the lattice, with vertices at $\pm \lambda_j$, while 16 are on layer 2 and so on. Then we have the estimate that this nth layer contribution to the sum is at most $8n \cdot (nR)^{-s}$, with $R = \inf\{|z| : z \in \partial P_1\}$.

Now, since $\sum_n n^{-s+1}$ converges for $s > 2$, we infer that G_s converges absolutely. □

Notice that the series $G_s(\lambda_1, \lambda_2)$ is homogeneous of degree $-s$, and that applying an element of $\mathrm{SL}(2, \mathbb{Z})$ to the lattice generators amounts simply to a rearrangement of the terms in the series, which implies the following statement.

Theorem 3.4. *Fix an even integer $2k \geq 4$.*

(a) *The series for $G_{2k}(\lambda_1, \lambda_2)$ defines an $\mathrm{SL}(2, \mathbb{Z})$-invariant function on $\mathcal{L}$ of weight k.*
(b) *The function $G_{2k}(z, 1)$ (denoted $G_{2k}(z)$) defines a modular form of weight k on the upper half-plane $\mathcal{U}$.*

Proof. The first statement (a) is clear from the definitions.

(b) Using the homogeneity, it follows immediately that the $\mathrm{SL}(2,\mathbb{Z})$-invariant function
$$G_{2k}(\lambda_1, \lambda_2) = \lambda_2^{-2k} G_{2k}(\lambda_1/\lambda_2, 1)$$
defines a function
$$G_{2k}(z) = \sideset{}{'}\sum_{\lambda \in \Lambda} 1/(mz+n)^{2k} \qquad (3.2)$$
on $\mathcal{U}$ which transforms as a modular form of weight k for $\Gamma(1)$.

We need to prove $G_{2k}(z)$ is holomorphic on $\mathcal{U}$ and at ∞; to do this we use an easy estimate of the summands on the fundamental set $\mathcal{D}$ for $\Gamma(1)$.

Lemma 3.5. *For $z \in \mathcal{D}$, we have*
$$|mz+n|^2 \geq m^2 - mn + n^2 = |m\rho - n|^2.$$
This is a routine computation.

It follows that the series G_{2k} converges uniformly and absolutely on $\mathcal{D}$ since $\sum 1/|m\rho - n|^{2k}$ does. The same is true on $\mathcal{U}$ too, since it holds for $G_{2k}(\gamma^{-1}z)$ in $\gamma\mathcal{D}$, for every $\gamma \in \Gamma(1)$, and the union of these sets covers $\mathcal{U}$. This implies that $G_{2k}(z)$ is holomorphic in $\mathcal{U}$.

The holomorphy at ∞ follows from the next result below, which concludes the proof of the theorem.

Lemma 3.6. *As $\Im(z) \to \infty$ in $\mathcal{D}$, $G_{2k}(z) \to 2\zeta(2k)$.*

Proof. Because of uniform convergence, we can pass to the limit as $\Im(z) = y \to \infty$ term by term: if $m \neq 0$ then terms $(mz+n)^{2k} \to 0$, while for $m = 0$ we get $1/n^{2k}$ and hence,
$$\lim_{\Im(z) \to \infty} G_{2k}(z) = \sideset{}{'}\sum_{n \in \mathbb{Z}} 1/n^{2k} = 2\zeta(2k). \qquad \square$$

Next, seeking a meromorphic Λ-periodic function in the plane, we observe a further consequence of Theorem 3.4.

Corollary 3.7. *The series*
$$\sum_{\lambda \in \Lambda} (z+\lambda)^{-s} \qquad (3.3)$$
converges uniformly and absolutely for z in compact subsets of $\mathbb{C}$ and for the range of values $s > 2$.

In this way, we obtain an infinite sequence of functions $F_n(z)$, $n \geq 3$, which are clearly holomorphic in z except for poles at the lattice points. These functions are also Λ-periodic: $F(z + \lambda) = F(z)$ for all $\lambda \in \Lambda$, simply by rearranging the absolutely convergent series.

The first function $F_3(z)$ of this sequence is, up to a constant multiple, the derivative of the famous Weierstrass $\wp$-function, the best-known of all elliptic functions.

3.3. Elliptic curves in Weierstrass form

Convergence properties of $\wp(z, \Lambda)$ incorporate modularity. We outline below very briefly how this story unfolds, the search for meromorphic functions on a complex torus $X(\Lambda)$ by way of functions on the entire plane which are Λ-periodic, usually termed *elliptic functions*.

Having decided that *there are no entire holomorphic functions with a given lattice of periods* because of Liouville's theorem, we move on to meromorphic functions.

Since any periodic meromorphic function takes every value in $\mathbb{C} \cup \infty$ a fixed number of times (see Exercises for Section 3), it follows that any such function with a single pole would determine a bijective isomorphism between $X(\Lambda)$ and $\widehat{\mathbb{C}}$, contradicting the fact that $\pi_1(X) \cong \mathbb{Z} + \mathbb{Z}$. A natural choice of location for a pole is the origin and we construct the Weierstrass functions, $\wp(z)$ and $\wp'(z)$ for a given lattice Λ, as functions holomorphic except for poles of order 2 and 3 at 0. Of course, periodicity implies that such functions will have a pole at every lattice point. The construction is made easier by our discussion of Eisenstein series.

Consider first the series for $\wp'$:

$$\wp'(z, \Lambda) = -2 \sum_{\lambda \in \Lambda} (z + \lambda)^{-3}. \tag{3.4}$$

This converges uniformly and absolutely on compact sets (after omitting finitely many polar terms) by comparison with $G_6(\Lambda)$ as we saw above.

To obtain $\wp(z)$, we integrate separately the polar term at 0 from $F_3(z)$ and integrate termwise $-2 \sum_{\lambda \in \Lambda^*}$ along the line segment from 0 to z. This uniformly and absolutely convergent series gives the (also uniformly absolutely convergent) series

$$\sum_{\lambda \in \Lambda^*} [(z - \lambda)^{-2} - \lambda^{-2}].$$

This produces the meromorphic function

$$\wp(z) = \frac{1}{z^2} + \sum_{\lambda \in \Lambda^*}[(z-\lambda)^{-2} - \lambda^{-2}]. \tag{3.5}$$

$\wp(z)$ is an even function (whereas $\wp'(z)$ is odd) but the fact that it is also Λ-periodic needs a little argument, which is left as an exercise for the reader, using the hint that for each $\lambda \in \Lambda$, both $\wp(z)$ and $\wp(z+\lambda)$ have $\wp'(z)$ as derivative, hence differ by a constant, to be shown to equal 0.

Next, we apply a simple trick to show that $\wp$ and $\wp'$ are algebraically related. Modulo the lattice, both have only one singularity, a pole at the origin, and suitable algebraic expressions in the two functions can be found which have the same higher order pole, whence they differ by a constant. We need to calculate their Laurent series.

Proposition 3.8. *The Weierstrass $\wp$-function has the following Laurent expansion at 0:*

$$\wp(z) = z^{-2} + \sum_{k=2}^{\infty}(2k-1)G_{2k}(\Lambda)z^{2k-2}. \tag{3.6}$$

For the proof, see Exercises for Section 3.

It follows that

$$\wp'(z) = -2z^{-3} + \sum_{k=2}^{\infty}(2k-1)(2k-2)G_{2k}(\Lambda)z^{2k-3}, \tag{3.7}$$

and so, working out the first few terms of the expression $\wp'^2 - 4\wp^3$, the series has only one polar term:

$$\wp'^2 - 4\wp^3 = -60G_4(\Lambda)z^{-2} + \text{ a power series in } z.$$

In fact it is easily checked that the series on the right side is of the form $-60G_4\wp(z) - 140G_6 + O(z^2)$, and the holomorphic part must be constant since the whole expression is Λ-periodic. This proves the fundamental *differential equation for $\wp(z)$*:

Theorem 3.9. *Let Λ be any plane lattice. The meromorphic Λ-periodic function $\wp(z,\Lambda)$ given by equation (3.5) satisfies the differential equation*

$$\wp'^2 = 4\wp(z)^3 - g_2\wp(z) - g_3, \tag{3.8}$$

where $g_2 = 60G_4(\Lambda)$ and $g_3 = 140G_6(\Lambda)$.

3.4. The discriminant form

In this section we employ w as lattice parameter in $\mathcal{U}$ to avoid confusion with $z \in \mathbb{C}$ as used in our account of elliptic functions in Section 2 (and continued here).

The discriminant form $\Delta(w) = g_2^3 - 27g_3^2$ is a member of the ring of modular forms, having weight 12. Its value at ∞ is equal to 0 from the determination in Lemma 3.6 of $G_4(\infty)$ and $G_6(\infty)$:

$$g_2(\infty) = 120\zeta(4) = 4\pi^4/3, \quad g_3(\infty) = 280\zeta(6) = 8\pi^6/27,$$

so Δ is a *cusp form*.

As a function of the lattice $L(w, 1)$ given by $w = \lambda_1/\lambda_2 \in \mathcal{U}$, $\Delta(w)$ is the squared product of the difference between the roots of the cubic expression in equation (3.8). Thus the question whether $\Delta(w) = 0$ in $\mathcal{U}$ comes down to the non-singular nature of the Weierstrass cubic.

We discuss briefly the geometric properties of the $\wp$-function viewed as a map on our elliptic curve $X = \mathbb{C}/\Lambda$. As an even meromorphic function, it determines a surjective holomorphic mapping $p : X \to \widehat{\mathbb{C}}$ and has degree 2 at every point (counting multiplicities), that is, every value in the Riemann sphere is covered twice except for branching, which occurs at the zeroes of $\wp'$ and at $0 = p^{-1}(\infty)$. It is not hard to see that these zeros are at the half periods $\lambda_i/2, \lambda_1+\lambda_2$. Noting that they are distinct — if not, then the degree of the map p would be 4 at least — we have the important implication that the three roots of the cubic expression in equation (3.8) are distinct. We have proved the following result.

Theorem 3.10. *The discriminant function $\Delta(w)$ is holomorphic and non-zero in $\mathcal{U}$.*

This has an important geometric consequence.

Corollary 3.11. *The mapping defined by the rule $z \mapsto (1, \wp(z), \wp'(z))$ induces a projective embedding of the elliptic curve $X(w, 1)$ as the non-singular homogeneous cubic curve $\mathcal{C}_w$ in $\mathbb{C}P^2$ given by*

$$Y^2 T = 4X^3 - g_2(w)XT^2 - g_3(w)T^3. \tag{3.9}$$

This mapping from $\mathbb{C}$ into $\mathbb{C}^3$ sends the orbit of 0 to the point $(0, 0, 1)$. For every lattice $\Lambda = L(w, 1)$, our discussion shows that the other points of $\mathcal{C}_w$ correspond bijectively to a unique point in $\mathbb{C}$ modulo $L(w, 1)$.

3.5. The space of modular forms and the j-invariant

Consider the set of all modular forms $\mathcal{M}_k$ of fixed weight $2k \geq 0, k \in \mathbb{Z}$: using the interpretation as rational differential forms of weight k on the modular surface, the Riemann sphere, we can infer that this is a finite-dimensional complex vector space, whose dimension depends on k.

The subspace of *cusp forms* $\mathcal{M}_k^0$ is the kernel of the linear morphism evaluation at ∞, $ev_\infty : f \mapsto f(\infty)$ from $\mathcal{M}_k$ to $\mathbb{C}$; we saw earlier that the Eisenstein series $G_{2k} \in \mathcal{M}_k$ has $ev_\infty(G_{2k}) \neq 0$. Therefore, we have

$$\mathcal{M}_k \cong \mathcal{M}_k^0 \oplus \mathbb{C}.G_{2k}, \quad \text{for each } k \geq 2.$$

Recall also that we have defined the discriminant $\Delta = g_2^3 - 27g_3^2$, an element of $\mathcal{M}_6^0$, where $g_2 = 60G_4$ and $g_3 = 140G_6$. We summarise below the dimensions of these spaces for small values of k.

Proposition 3.12.

(a) $\mathcal{M}_k = 0$ if $k < 0$ or $k = 1$.
(b) $\mathcal{M}_0 = \mathbb{C}$. If $2 \leq k \leq 5$, $\mathcal{M}_k = \mathbb{C}.G_{2k}$; there are no cusp forms of weight $k < 6$.
(c) The rule $f \mapsto f.\Delta$ defines an isomorphism $\mathcal{M}_k \cong \mathcal{M}_{k+6}^0$ for each k.

These are all simple consequences of the basic equation below which expresses the divisor of a modular form $f \in \mathcal{M}_k$ in terms of its order of vanishing at the three special points of $\widehat{X(1)}$:

$$\nu_\infty(f) + \frac{1}{2}\nu_i(f) + \frac{1}{3}\nu_\rho(f) + \sum_{p \in \mathcal{U}/\Gamma(1)} \nu_p(f) = \frac{k}{6}. \tag{3.10}$$

The equation results from expressing the divisor of the rational form $f(z)dz^k$ on $X(1)$ in terms of local coordinates at ∞ and at the key points i, ρ of $\mathcal{U}$. It can be proved using the Riemann–Roch theorem or by direct contour integration of df/f around the frontier of the fundamental polygon $\mathcal{D}$ with local detours to avoid the special points. See [28] for details.

Corollary 3.13. *For all $k \geq 0$,*

$$\dim \mathcal{M}_k = \begin{cases} [k/6] & \text{when } k \equiv 1 \pmod{6}, \\ [k/6] + 1 & \text{when } k \not\equiv 1 \pmod{6}. \end{cases}$$

This is proved by induction on k or deduced from the Riemann–Roch theorem for $\widehat{X(1)}$, which inter-relates the dimensions of spaces of rational functions and forms on $\widehat{\mathbb{C}}$. See for instance [7].

To define the most important modular invariant, $j(w)$ for $w \in \mathcal{U}$, we write Λ_w for the lattice $\Lambda(w, 1)$, and $g_2(w)$, $g_3(w)$ and $\Delta(w)$ for the corresponding functions of Λ_w. The two functions g_2^3 and g_3^2 are clearly linearly independent modular forms of weight 6 not in $\mathcal{M}_6^0$. Define

$$j(w) = 1728 g_2^3(w)/\Delta(w). \qquad (3.11)$$

By calculating a few terms in the q-expansions of g_2 and g_3, it can easily be shown that

Proposition 3.14. Δ *has a simple zero at* ∞.

The constant is chosen so as to arrange for the Laurent series for $j(w)$ at ∞ to begin with $1/q + \cdots$. Since we know that $\Delta(z) \neq 0$ in $\mathcal{U}$, we have the following proposition.

Proposition 3.15. *The j-function defines an isomorphism between* $\widehat{X(1)}$ *and* $\widehat{\mathbb{C}}$.

A more careful analysis (see [7, 33]) shows that Δ and j have integer coefficients of great arithmetic interest.

Theorem 3.16. $\Delta(q) = (2\pi)^{12} q(1 + a_1 q + \cdots)$ *and* $j(q) = q^{-1} + 744 + b_1 q + \cdots$, *with rational integer coefficients.*

The product expression given in Section 2.5 to illustrate the concept of modular form turns out to coincide with the form $\Delta(q)$, up to a multiplicative constant. This follows from a tricky calculation (see [28]) showing it is a modular form of weight 6: properties outlined above for the algebra of modular forms then imply the result. The coefficients a_n define the Ramanujan τ-function, $\tau(n) = a_n$ and form the subject of his famous conjecture, generalised by H. Petersson to all suitably normalised cusp forms which have a certain kind of Euler product formula (again, see [28]); this was proved by P. Deligne in 1973 using his proof of the Weil conjectures.

Note. Many authors use a different normalisation for the Eisenstein series, using instead the functions $E_n(z) = G_n(z)/G_n(\infty)$ which take the value 1 at ∞. This has distinct advantages from the arithmetic point of view.

3.6. *Exercises for Section 3*

(1) Show using covering space theory that two lattices Λ, Λ' are similar if and only if the corresponding complex tori $X = \mathbb{C}/\Lambda$, $X' = \mathbb{C}/\Lambda'$ are biholomorphically equivalent.

(2) Calculate the Laurent series expansion of the function

$$\wp(z, \Lambda) = z^{-2} + \sum_{n=0}^{\infty} c_n z^n$$

in (even) powers of z and use it to calculate the Laurent series of the Weierstrass function $\wp'(z)$ and check the validity of the differential equation

$$\wp'(z, \Lambda)^2 = 4\wp^3 - g_2(\Lambda)\wp - g_3(\Lambda).$$

Use the differential equation to show that each coefficient in the expansion of $\wp$ is a polynomial in the two coefficients g_2 and g_3.

(3) Show that the lattice $L(\rho) = \mathbb{Z} \oplus \rho\mathbb{Z}$, where $\rho = e^{\pi i/3}$, is preserved by Euclidean rotation through angle $\pi/3$ and deduce that the torus $X(\rho) = \mathbb{C}/L(\rho)$ has an order 6 conformal automorphism. Verify that the Eisenstein series $G_6(w)$ vanishes at $w = \rho$.

Work out the corresponding facts for the square lattice $L(i)$.

(4) Let f be an elliptic function for a lattice Λ. By integrating the 1-form $d \log f(z)$ around the edge of a fundamental polygon Π, show that f has an equal number of zeros and poles on the torus $X = \mathbb{C}/\Lambda$. Deduce that f takes on every value the same number of times on X; this integer is called the *degree of* f.

Use a similar argument to show that the degree of an elliptic function cannot be 1.

(5) Show that each space $\mathcal{M}_k$ is spanned by monomials $g_2^m g_3^n$ with $2m + 3n = k$. [*Hint*: prove it by induction on k, using properties of Δ.]

(6) Let f be a modular function. Show that f is a rational function of $j(z)$.

(7) (*The tree of* $\mathrm{SL}(2, \mathbb{Z})$.) Consider the geodesic segment $\ell \subset \mathcal{U}$ which joins the points 1 and ρ.

  (a) Show that $j(\ell)$ is the real line segment $0 \le x \le 1728$.
  (b) Show that the set of all points $\gamma(z)$ with $\gamma \in \Gamma(1)$ and $z \in \ell$ is a tree $\mathcal{T} \subset \mathcal{U}$, i.e., a connected graph without circuits, on which the action of Γ on $\mathcal{T}$ is discontinuous with fundamental set ℓ.

Note: It follows from part (b) and the Bass–Serre theory of groups acting on trees (see [29, p. 35]) that the group $\mathrm{PSL}(2, \mathbb{Z})$ is isomorphic to the free product of the cyclic groups $C_2 \cong \mathrm{Stab}(i)$ and $C_3 \cong \mathrm{Stab}(\rho)$.

4. Horocycles, Ford Circles and Farey Fractions

Within the environment of conformal geometry in the plane, there is a particular pattern of circles relevant to the modular group action which is closely linked to an inductive construction of the rational numbers familiar from elementary number theory and also to the classical study of continued fractions.

We recall first some background facts of conformal plane geometry which will be important. Proofs are elementary and left to the reader.

4.1. *Circle geometry and horocycles of* $\Gamma(1)$

We saw that the modular group $\Gamma(1)$ acts discretely by fractional linear (Möbius) transformations on the upper half-plane, to produce a quotient surface isomorphic to the complex plane, which is completed by the addition of a point at infinity to yield the *Gauss–Riemann sphere*, $\widehat{\mathbb{C}}$ — this space is familiar to topologists as the 2-sphere, and to algebraic geometers as the *complex projective line* $\mathbb{CP}^1$. Furthermore it represents the most basic instance of how a compact Riemann surface carries within itself a natural geometric structure.

The complex Möbius group $\mathrm{PSL}(2,\mathbb{C})$, also known as the *2D conformal group*, plays a crucial role in low-dimensional geometry and in conformal field theory. It comprises the biholomorphic automorphisms of the Riemann sphere, and the action preserves not only the purely local notion of angle but also the set of Euclidean lines and circles. And this is not the end of the catalogue of interesting features: for each Euclidean circle/line, there is an associated stability subgroup, conjugate to the real Möbius group, $\mathrm{PSL}(2,\mathbb{R})$. The real Möbius group has its own storehouse of geometric treasures as we have seen: it leaves invariant the completed real line $\widehat{\mathbb{R}} = \mathbb{R} \cup \infty$, the upper half-plane $\mathcal{U}$, the hyperbolic metric on $\mathcal{U}$ and the set of hyperbolic geodesic lines.

The *rational boundary points* $\widehat{\mathbb{Q}} \subset \widehat{\mathbb{R}}$ form a single orbit under the modular group $\mathrm{PSL}(2,\mathbb{Z})$ and each point $x = p/q \in \mathbb{Q}$ has stability subgroup conjugate to the (translation) subgroup $\langle T \rangle$ fixing the point at infinity.

Now we consider the family of all circles lying in the closure of the upper half-plane $\mathcal{U}$ which are tangent to the boundary real line/circle: for the point at infinity, we use the horizontal lines in $\mathcal{U}$, parallel to the boundary. We may view these as circles too by adding the extra point ∞. All of these tangent circles are known as *horocycles*. Just as the set of horizontal lines is preserved by the translation subgroup $\langle T \rangle$, so it follows that each horocycle is preserved (as a set) by a conjugate stability subgroup of parabolic elements fixing the point of tangency.

As we know, the involution U together with translation T generates $\Gamma(1)$. Therefore, we can understand the effect of an arbitrary element of $\Gamma(1)$ on horocycles by applying these two repeatedly. Thus, horocycles at infinity are sent by U to horocycles at 0, whence they are translated by T to arbitrary integer points. A second application of U then sends these to horocycles based at the points $1/n$, with $n \neq 0$, and so on.

4.2. Ford circles

Let p/q be a rational fraction in reduced form, with $q > 0$.

Definition 4.1. The *Ford circle* $C(p/q)$ *based at* p/q is a horocycle of specified radius $1/(2q^2)$: thus, it is the unique circle of this radius in the upper half-plane tangent to the real line at p/q.

We add to this set a further member based at ∞: the *Ford circle* $C(1/0)$ *based at* ∞ is defined to be the completed line $\Im(z) = 1$.

The properties of this family of circles are best displayed by a picture, which indicates that *the set of Ford circles forms a (rigid) pattern consisting of either mutually tangent or disjoint circles* — a simple illustration is given in Fig. 2 which omits the horizontal line $C(1/0)$ touching all the circles $C(n/1)$, $n \in \mathbb{Z}$.

This picture represents the first few circles lying between $C(0/1)$ and $C(1/1)$ with p/q in the unit interval; translation $T : p/q \to (p+q)/q$ operates naturally as a symmetry of the complete set, and the figure shows some circles within a fundamental set for this cyclic subgroup.

Other images are readily found under *Ford circles* on the WWWeb.

The next result asserts that the modular group is directly involved.

Proposition 4.2. *The set of Ford circles forms a single orbit under the action of the modular group* $\Gamma(1)$ *on the space of horocycles.*

The proof is a simple exercise, using only what we know about generators of $\Gamma(1)$ acting on the orbit of ∞.

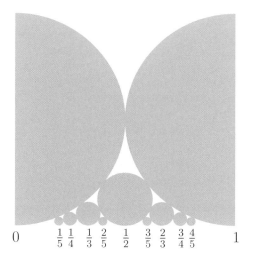

Fig. 2. Some Ford circles.

4.3. Farey sequences

An old approach to enumerating the set of all rational fractions inductively was described in a letter to the Philosophical Magazine in 1816 by a geologist named John Farey; this idea was then justified and popularised by A. Cauchy. It goes as follows: restricting attention to the unit interval and the set of p/q with $0 = 0/1 \leq p/q \leq 1/1 = 1$ for simplicity, consider for each integer $n > 0$ the ordered list of fractions with denominator at most n. Thus we get at stage 3 the finite sequence

$$0 = 0/1 < 1/3 < 1/2 < 2/3 < 1/1.$$

Each successive stage involves insertion of the fractions p/q with $\gcd(p,q) = 1$ and $q = n$: no denominator greater than n is included. Clearly, none of the stage n list can be already in the stage $n-1$ list, so this provides an enumeration (with no repetitions) of all rationals in the interval.

Definition 4.3. Two rationals p/q, r/s are called *Farey neighbours* if they satisfy the condition

$$ps - rq = \pm 1. \tag{4.1}$$

Notice that for this to hold the rationals must be in reduced form, i.e., $\gcd(p,q) = \gcd(r,s) = 1$.

One verifies easily that this neighbour condition expresses a general homogeneity property of the set of all Farey pairs under the action of the *extended modular group* $\mathrm{GL}(2, \mathbb{Z})$, stated as follows.

Proposition 4.4. *Two rationals p/q, r/s are Farey neighbours if and only if the two points are the images of the pair 0, ∞ respectively under the integer Möbius map*

$$z \mapsto \frac{pz + r}{qz + s}.$$

Definition 4.5. The *Farey mediant* of $p/q, r/s$ is the fraction $(p+r)/(q+s)$.

See Fig. 3 which shows how this relationship determines the basic pattern of Ford circles.

Notice that passing from one Farey sequence to the next involves the insertion of all Farey mediants with denominator at most n. This involves only finitely many terms at each stage within any finite interval.

For instance, at the beginning of a sequence in the unit interval one needs to insert

$$0/1 < 1/n < 1/(n-1) < \cdots < 1/2 < \cdots < 1.$$

The key fact about Farey sequences is given in the next statement, which follows immediately from the definition.

Proposition 4.6. *Adjacent terms in each Farey sequence have the Farey neighbour property.*

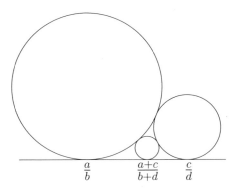

Fig. 3. A mediant circle.

A simple procedure to generate the next Farey sequence at each level iteratively is explained in [25]: it involves only the formation of all mediants from the previous level together with rejection of those for which the denominator is too large. This produces a proof of the following fact.

Proposition 4.7. *Let $p < q$ be relatively prime natural numbers. Then p/q is a reduced fraction which appears in the Farey sequence of level q. In this sequence, it will have two Farey neighbours given by*

$$h_1/k_1 < p/q < h_2/k_2,$$

with $qh_1 - pk_1 = -1$ and $qh_2 - pk_2 = 1$.

Furthermore, p/q is the mediant of these two fractions, that is, we have $p = h_1 + h_2$ and $q = k_1 + k_2$.

Thus, we may conclude the following fact.

Corollary 4.8. *There are infinitely many Farey neighbours for each rational p/q.*

Proof. Apply Proposition 4.4, noting that the equation $px - qy = 1$ has infinitely many solutions in integers if p and q are coprime. $\square$

4.4. Farey sequences and Ford circles

There is a beautiful link between these notions, which is treated in [25]. Consider the set of three Ford circles based at $\infty = 1/0$, $0/1$ and $1/1$ respectively: one sees easily that they are mutually tangent. Notice also that $1/1$ is the Farey mediant of the other two: here we need to stretch our definitions a little — we have only described the pattern for points in the unit interval, although it extends immediately to the whole line $\mathbb{R}$ by the homogeneity property above and completes to infinity with the rule that $\infty = 1/0$ together with inclusion of the fraction $1/0$ in the definition of Farey neighbours.

This link to Ford circles then holds true in general: in other words, we have the following theorem.

Theorem 4.9. *The collection of all Ford circles has the property that two circles are mutually tangent if and only if the corresponding base points are Farey neighbours.*

For the proof, we only need to observe that (i) the statement holds for two points/circles, and (ii) both properties are invariant under application of any element of the modular group.

4.5. Diophantine approximation of irrational numbers

A simple observation shows how to interpret the Ford circles configuration to see that there are limits to how well one can approximate an irrational by rational numbers.

Consider a vertical line in the upper half-plane through an irrational number, $\ell_x = \{\Re(z) = x\}$. Because x is not rational, there is no circle from the Ford system tangent at x. Furthermore, any Ford circle which intersects ℓ_x is surrounded below by other circles, so that ℓ_x must enter a lower circle. Hence, by repeating this step it meets infinitely many circles $C(p/q)$ of the system on the way to x. This implies the following theorem.

Theorem 4.10. *For every irrational x, there is an infinite sequence of rationals p/q, such that*

$$|x - p/q| < 1/2q^2. \qquad (4.2)$$

With a more careful analysis of the situation, one can prove using the pattern of Ford circles the following stronger result of A. Hurwitz on approximation of irrational x.

Theorem 4.11. *For any irrational number x, there exist infinitely many fractions p/q such that*

$$\left| x - \frac{p}{q} \right| < \frac{1}{\sqrt{5}q^2}.$$

See [25] for details and for an explanation why this result of Hurwitz is *optimal* in the sense that no smaller constant than $1/\sqrt{5}$ will serve in this way.

The classical approach to Diophantine approximation uses the technique of *continued fractions* pioneered by Euler; see for instance [12]. These are a sequence of rational approximants to a real number x, beginning with $q_0 = [x]$, the integer part of x, for which we see that $x = q_0 + 1/x_0$ with $x_0 > 1$. Repeating this simple step produces an expression

$$x = q_0 + \frac{1}{q_1 + 1/x_1}, \quad \text{with } x_1 > 1.$$

Continuing the process produces expressions with repeated fractions which we write as follows:

$$q_0 + \frac{1}{q_1+} \frac{1}{q_2+} \frac{1}{q_3 + 1/x_3}$$

with $x_3 > 1$, which represents a better approximation to x.

Note that if $x = p/q$ is a rational number in lowest terms, then the process is essentially a repeated application of the Euclidean algorithm, terminating in finitely many stages. For instance,

$$\frac{17}{11} = 1 + \frac{1}{1+} \frac{1}{1+} \frac{1}{5}.$$

If x is irrational, then the sequence is infinite and the question of convergence arises. Writing $\frac{A_n}{B_n}$ for the rational number obtained by omitting the final term $1/x_n$, we obtain the sequence of *convergents*, which satisfies the property

$$x - \frac{A_n}{B_n} = \frac{(-1)^n}{B_n(x_n B_n + B_{n-1})}, \qquad (4.3)$$

from which convergence to x follows easily. In fact, it is not hard to show that each convergent is related to its predecessor by the familiar relation

$$A_n B_{n-1} - A_{n-1} B_n = (-1)^{n-1},$$

providing an immediate connection to Farey neighbours, see equation (4.1).

4.6. The Farey graph

The collection of fractions $\mathbb{Q} \cup \infty$ written p/q and $1/0$ in lowest terms is made into a graph in the upper half-plane by joining each pair of Farey neighbours by the hyperbolic geodesic for which they are the boundary points. We note the following fact.

Proposition 4.12. *Each triple of rationals which consists of two Farey neighbours p/q, r/s and their mediant $(p + r)/(q + s)$ forms the vertices of an ideal triangle with the property that no other Farey edge intersects it.*

This follows at once from earlier results about the Farey numbers as a method to exhaust the rationals.

The procedure creates a tiling of $\mathcal{U}$ by ideal triangles, called the *Farey tessellation*. The 1-skeleton in the partially compactified disc $\mathcal{U} \cup \widehat{\mathbb{Q}}$ is called the *Farey graph* $\mathcal{F}$. This has played an important part in the modern theory of Kleinian groups, where a key initial approach to Thurston's work on hyperbolic 3-manifolds has been the study of *punctured tori* and the classification of discrete subgroups of $\mathrm{PSL}(2, \mathbb{C})$ which represent them; see [24].

4.7. Exercises for Section 4

(1) Find all Farey neighbours of the points $0/1$ and $1/1$. Draw the family of geodesic edges of $\mathcal{F}$ so defined.

(2) Draw a diagram of (a small part of) the Farey tessellation. Find a fundamental domain for the action of $\Gamma(1)$ on this ideal triangulation of $\mathcal{U}$.

  How does the set of Ford circles fit into your Farey graph picture?

(3) Show that for any two tangent Ford circles there is a unique third circle tangent to the other two which touches the boundary at the mediant rational point. Find the points of tangency of these circles: show they have always a rational x-coordinate.

(4) *Continued Fractions.* Using this classical approach to approximation of real numbers, as outlined in Section 4.6, show that the *rational approximants* A_n/B_n to a real number satisfy the Farey neighbour property, equation (4.1).

  Note. It is an interesting exercise to investigate the case of a *quadratic irrational* $x = \sqrt{N}$, which involves discussion of integer solutions to Pell's equation; see Davenport's book [12] for instance.

(5) Use the hyperbolic reflection group Γ generated by reflections in the three geodesics which connect the vertices $1/0$, $0/1$ and $1/1$ in the Farey graph to generate the entire tessellation. What does this imply for the continued fraction process?

5. Topics in Automorphic Forms

This section begins by outlining a systematic approach to the study of modular and automorphic forms. A wider goal is to indicate how the method is applicable for the study of any finite-type (hyperbolic) Riemann surface, although there is no opportunity to pursue it here.

5.1. *Historical remarks*

The idea of an automorphic function has origins in Gauss's study of the hypergeometric equation and the pioneering work of H.A. Schwartz on *triangle groups and functions*. Consider a triangle with sides which are arcs of circles or line segments and with angles of the form $\pi/l, \pi/m, \pi/n$ such that

$$\frac{\pi}{l} + \frac{\pi}{m} + \frac{\pi}{n} < 1 \tag{5.1}$$

determines a hyperbolic *tessellation of a disc* or half-plane, by the procedure of reflecting the triangle in each side and repeating this activity. See for instance the books [9, 11]. As an important example of this construction, the triangle which is formed by bisecting $\mathcal{D}$ along its imaginary axis of symmetry has angles $\pi/2, \pi/3$ and $\pi/\infty = 0$ and the above process recreates the modular group tessellation of $\mathcal{U}$.

Using techniques from geometric function theory, we can deduce the existence of a certain automorphic function for the triangle group $\Gamma(l, m, n)$ produced by this reflection process in $\mathcal{U}$. For more details, the reader may consult the textbook [9] or [23].

The *Riemann mapping theorem* states that there is a bijective holomorphic map from the interior of the triangle $T_{l,m,n}$ to the unit disc (or upper half-plane, as these spaces are isomorphic). This map extends continuously to a map of the triangle's boundary onto the circle (or completed real axis) and if we specify boundary values at three points, say $f(i) = 0$, $f(\rho) = 1$ $f(\infty) = \infty$ for $f : T_{l,m,n} \to \mathcal{U}$, the function is unique. The *Schwartz reflection principle* shows that this function extends to a surjective biholomorphic map from the union of triangles and conjugates, i.e., $\mathcal{U}$, into $\widehat{\mathbb{C}}$, the Riemann sphere; this construction gives the *j*-function (up to a constant multiple).

Poincaré and Klein studied automorphic forms for arbitrary Fuchsian (finitely generated) discrete groups $\Gamma \subset \text{PSL}(2, \mathbb{R})$ in the late 19th Century, establishing the groundwork for a complete understanding of the basic structure of fields of *automorphic functions*, which are meromorphic functions and vector spaces of forms on the Riemann surfaces $X(\Gamma) = \mathcal{U}/\Gamma$. Besides laying the foundations for the contemporary theory of Teichmüller spaces and applications in geometry of 3-manifolds, this has developed into a vast area encompassing the representation theory of general lattices in Lie groups and inspired an elaborate conjectural unifying picture of many aspects of mathematics (including number theory, arithmetic geometry, conformal field theory, etc.) within a framework loosely called the *Langlands Programme*.

5.2. *Finite type and finite area surfaces*

Finiteness conditions are important in the general study of Riemann surfaces as they provide a framework in which a systematic approach can be made to the analytic theory. In the present setting, we have the opportunity afforded by the underlying hyperbolic geometry of the upper half-plane in which two notions of finiteness for a quotient surface turn out to be equivalent. As Poincaré saw in a moment of inspiration, the non-Euclidean geometry in $\mathcal{U}$ is inherited by every Riemann surface covered by $\mathcal{U}$: this includes all the closed surfaces of genus 2 or more as well as modular surfaces. Any surface covered by $\mathcal{U}$ can be reconstructed geometrically, using polygons which tessellate the hyperbolic plane in the same way that parallelograms tile the Euclidean plane. In fact, as a consequence of the Koebe–Poincaré Uniformisation Theorem, this construction method applies without exception to *all Riemann surfaces with negative Euler characteristic*.

Definition 5.1. A Riemann surface X is said to be *of finite type* if it has finite topological type, that is, provided that the Euler characteristic $\chi(X)$ is finite.

Now it turns out that whenever a surface is expressible as a quotient space $X = \mathcal{U}/\Gamma$ of Γ-orbits for a (finitely generated) discrete group Γ-action in the hyperbolic plane, there is a direct geometric construction for X analogous to the process described in Section 2 for the modular group, with passage to a quotient space $\mathcal{P}/\sim$ given by identifying pairs of sides of a polygonal fundamental polygon $\mathcal{P}$, and resulting in a finite number of special orbits representing cone points or cusps. There may also be a finite number of open boundary curves if the quotient surface has infinite area.

Note 1. For Γ a subgroup of the modular group of course, any cusp point Q lies in $\mathbb{Q} \cup \infty$ and the stabiliser is conjugate to $\langle z \mapsto z + b \rangle$ with $b \in \mathbb{Z}$. As we saw in Section 2, $\Gamma(1)$ itself has two orbits of cone points and one cusp (puncture), but finite index subgroups may have any number of cone points and cusps.

Note 2. There is an analogue of cusp neighbourhood for all elliptic fixed points $\zeta \in \mathcal{U}$: there exists an open set $V \subset \mathcal{U}$, a nbd of ζ with finite cyclic

Γ-stabiliser $\operatorname{Stab}_\Gamma(\zeta) = H = \langle \gamma | \gamma^m = 1 \rangle$, such that V is H-stable and projection $p: V \to V/H$ is m-to-1 onto a nbd of the image point $Q \in X$.

We can now complete discussion of finiteness for surfaces covered by the upper half-plane. For any cusp nbd, the contribution to the area is finite, even though it is non-compact, and if the quotient has finite area, then after cutting out a finite number of cusp nbds, the remainder is compact, i.e., a fundamental set is closed and bounded in $\mathcal{U}$, the result of truncating a convex Dirichlet polygon by removing finitely many horodiscs.

Theorem 5.2. *Finite area for the fundamental domain of a Fuchsian group Γ implies finite type for the Riemann surface $X = \mathcal{U}/\Gamma$.*

Conversely, any finite type hyperbolic Riemann surface has such a representation, by constructing a finite area fundamental domain for its fundamental group acting as covering isometries in the universal cover.

Proof. The area of a region in the hyperbolic plane is defined by integrating the *hyperbolic area element* $ds_h \times ds_h$, obtained as the (tensor product) square of the length element):

$$\operatorname{Area}(F) = \iint_F \frac{dx\,dy}{y^2}.$$

We consider a convex polygonal region F bounded by a finite number of hyperbolic line segments. To evaluate such an integral, we apply *Green's Theorem*:

$$\iint_F \left(\frac{\partial Q}{\partial x} - \frac{\partial P}{\partial y} \right) dx\,dy = \int_{\partial F} P\,dx + Q\,dy;$$

putting $P = 1/y$, $Q = 0$ we have $\partial P/\partial y = -1/y^2$, and so

$$\operatorname{Area}(F) = \int_{\partial F} \frac{dx}{y}.$$

Now this integral is easy to evaluate over any segment of hyperbolic geodesic in $\mathcal{U}$: if the path ℓ follows a hyperbolic line, a circular path of Euclidean radius r, from angle β to γ, then we find

$$\int_\ell \frac{dx}{y} = \int \frac{d(r\cos\theta)}{r\sin\theta} = -\int_\beta^\gamma d\theta = \beta - \gamma;$$

thus the answer is independent of r. $\square$

Exercise. Show that the integral along a vertical line segment is always zero.

Now it is easy to complete the calculation for a compact piecewise geodesic boundary ∂F: the total effect from a sequence of arcs $\ell_j, j = 1, \ldots, m$ is to sum the various angular differences

$$\text{Area}(F) = \sum_{j=1}^{m}(\beta_j - \gamma_j).$$

To simplify this further, we convert the sum into a sum of interior angles of the polygon F, using the outward pointing normals for the sides to keep track of our orientation as we turn the corners at each vertex. After one circuit of ∂dF we have turned through a total angle 2π, as the result of a change in the normal direction of $\gamma_j - \beta_j$ on each edge ℓ_j and $\pi - \alpha_k$ at each corner V_k so that

$$2\pi = m\pi - \sum_k \alpha_k + \sum_j (\gamma_j - \beta_j).$$

But now, putting this into our formula for Area(F), we deduce the *Gauss–Bonnet formula*.

Theorem 5.3. *For a hyperbolic polygon with m sides the hyperbolic area is given by the formula:*

$$\text{Area}(F) = (m-2)\pi - \sum_{k=1}^{m} \alpha_k.$$

This result embodies the difference between Euclidean and hyperbolic geometry. It makes explicit the effect of the *negative curvature* of hyperbolic space, contrasting with the formula for area in Euclidean or spherical geometry.

As a special case we highlight the case of a *triangle* Δ with angles α_k:

$$\text{Area}(\Delta) = \pi - (\alpha_1 + \alpha_2 + \alpha_3).$$

Corollary 5.4. *The area of the modular surface $X(1)$ is $\pi/3$.*

Proof. $\text{Area}(\mathcal{D}) = 2\pi(1 - \frac{1}{2} - \frac{1}{3})$. □

Corollary 5.5. *Let $\Gamma < \Gamma(1)$ be a subgroup of finite index m. Then $X = \mathcal{U}/\Gamma$ has area $m\pi/3$.*

Proof. There is a finite set of elements $\gamma_1, \ldots, \gamma_m$ in $\Gamma(1)$ which are distinct coset representatives for Γ. It follows easily that the union $\bigcup_{n=1}^{m} \gamma_n \mathcal{D}$ is (the interior of) a fundamental set for Γ which is a union of polygonal sets with total area $m\pi/3$. But one can show, in the same way as for lattices in the plane, that the area of any fundamental set for a Fuchsian group Γ is the same (see Exercises for this section or [5]). □

5.3. The space of lattices: A trefoil knot complement

We have seen how to exploit the function theory of Eisenstein series G_4 and G_6 in constructing other modular forms and functions. There is a beautiful topological spinoff from this material (see [22, p. 84], where it is attributed to D. Quillen) which ties it in with algebraic singularities and relates the trefoil knot complement to the modular group action on lattices.

Theorem 5.6. *The homogeneous space $H^3 = \mathrm{SL}(2,\mathbb{R})/\mathrm{SL}(2,\mathbb{Z})$ is diffeomorphic to the trefoil knot complement in the 3-sphere S^3.*

The proof goes as follows: H^3 is the space of *unit area* lattices in the plane, since $\mathrm{SL}(2,\mathbb{R})$ acts transitively on this space and $\mathrm{SL}(2,\mathbb{Z})$ stabilises the square lattice.

Recall that we saw $\mathcal{L} \cong M/\mathrm{SL}(2,\mathbb{Z})$ in Section 1.4, and we associated to each lattice L an elliptic curve $X(L) = \mathbb{C}/L$. We also defined the Weierstrass function $\wp(z, L)$ and saw in Section 3 that the differential equation $(\wp')^2 = 4\wp^3 - g_2\wp + g_3$ holds for the elliptic curve $X(L)$. For each lattice L, this cubic polynomial has three distinct roots in $\mathbb{C}$ because the curve is a complex torus: in fact the mapping $z \mapsto (\wp(z), \wp'(z))$ defines a surjective covering projection from $\mathbb{C} - L$ to $X(L) - \{\infty\}$.

Next, recall that the existence of multiple roots for this cubic is detected by vanishing of the *discriminant* which is given by $\Delta(L) = g_2^3 - 27g_3^2$. It is known, from the theory of differential equations for instance, that the pair $g_2(L), g_3(L)$ characterise the solution function $\wp(z, L)$ uniquely, and all pairs with $\Delta \neq 0$ occur. It follows that the rule which sends each lattice L to the pair $u = g_2(L), v = g_3(L)$ induces a map of H^3 to the complement of the vanishing locus given by the affine curve $u^3 - 27v^2 = 0$ in $\mathbb{C}^2$, which is a hypersurface with a singularity at the origin.

Following a standard method to analyse such singularities, the intersection of this locus with the unit 3-sphere in $\mathbb{C}^2$ is a trefoil knot complement.

The argument is completed by showing that for each lattice, there is a unique scaled lattice whose image lies on the unit sphere $|u|^2+|v|^2 = 1$. The reader should determine how the scaling factor $\lambda \in \mathbb{R}^*_{>0}$ involved depends on the lattice L.

Note. It is worth recalling here that the trefoil knot is homeomorphic to a $(2,3)$-torus knot $K_{2,3}$, a simple loop on a torus $S^1 \times S^1$ in $\mathbb{R}^3$ which winds twice around one generating circle while winding three times in the other. The fundamental group of $S^3 \setminus K_{2,3}$ is *Artin's braid group* B_3, isomorphic to the abstract group

$$\langle s, u | s^3 = u^2 \rangle.$$

See [23] and Exercises for Section 5.

5.4. *Conformal field theory and Moonshine*

The set of coefficients b_n in the q-expansion of $j(z)$ have an astonishing link to the largest sporadic finite simple group $\mathbb{M}$.

Following the original observation by John McKay that $b_1 = 196884$ is one more than the dimension of the fundamental representation of $\mathbb{M}$ (which at the time was only conjectured to exist), it was found that each b_n is a linear combination of characters of $\mathbb{M}$. This triggered the insights of Conway and Norton and further input by J.G. Thompson (and perhaps others) then spawned the *Moonshine conjectures* which link all conjugacy classes of elements of $\mathbb{M}$ to genus 0 modular functions representing some aspect of the group theory. See [10] for details.

Another indirect link to modular forms was Ogg's observation that the set of primes dividing the order of $\mathbb{M}$ coincides with the primes for which the (extended) Hecke group $\Gamma_0(p)+$ has quotient $X_0^+(p)$ of genus 0. A clear reason for this has never been given without invoking the full Moonshine.

The proof of the Moonshine conjectures by Borcherds involved the earlier construction (by Frenkel, Lepowsky and Meurman [14]) of a vertex operator algebra for which $\mathbb{M}$ is the automorphism group and further significant use of insights from conformal field theory to study these algebras in depth. See Borcherds' paper [6] for more discussion.

After reading Terry Gannon's detailed progress report from 2006 (see [15]), one is left with the feeling that we are little wiser ten years after the proof as to the true meaning of this mysterious connection between finite group theory, physics and the modular group.

5.5. Sundry topics

(1) *Modular elliptic curves and Fermat's last theorem.* The key ideas which inter-relate these concepts are outlined briefly, together with a sketch of the proof, in two exposés (numbers 803 by J.-P. Serre and 804 by J. Oesterlé) in Séminaire Bourbaki (Juin 1995).

The arithmetic theory of elliptic curves has not been addressed at all here, due to lack of space; along with the build-up to the conjecture of Taniyama–Weil which encapsulates much of the theory, it is well-described in [20]. A more elementary account can be found in [18], as well as in various popular math texts.

(2) *More number-theoretic applications*: see [37] by Don Zagier for further results by arithmetic methods, such as Hecke operators and Ramanujan's τ-function. See also [8] for a discussion of the part that modular forms played in Ramanujan's work on approximations to π.

(3) *Grothendieck's dessins d'enfants.* Finite index subgroups Γ of $\Gamma(1)$ determine Riemann surfaces $X(\Gamma)$ covering the modular surface $X(1)$ which are ramified over three points, two cone and one cusp: according to a theorem of A. Weil, these surfaces represent algebraic curves defined over a number field, and the converse statement is Belyi's theorem, which famously disconcerted both Grothendieck and Deligne: see [17].

This leads into Grothendieck's theory of dessins d'enfant which has generated an avalanche of articles in recent times. For the main ideas of this theory, see for instance [16, 30].

(4) *The Farey triangulation of $\mathcal{U}$.* Many applications of this simple but influential structure have appeared in the literature, since W.P. Thurston's revolutionary work on the part that hyperbolic geometry plays in the topology of 3-manifolds. See [3, 21, 34], and also [24] for a striking application resolving one case of Thurston's Ending Lamination conjecture in the area of Kleinian groups. A paper by A. Verjovsky [38] provides a link to continuing work on the Riemann Hypothesis.

5.6. Final exercises

(1) Prove that two triangles in hyperbolic geometry are *congruent* (i.e., isometrically equivalent) if and only if they have equal angles.
(2) Let $\Gamma(N)$ denote the *principal congruence subgroup* of level N, which is the kernel of the surjection from $\Gamma(1)$ to $\mathrm{SL}(2, \mathbb{Z}/N\mathbb{Z})$. Calculate the index (and hence the hyperbolic area of $X(N)$) for the cases $N = 2, 3$.

In the case $N = 2$, show that the stabiliser of ∞ in the subgroup is generated by $T^2 : z \mapsto z + 2$. Show also that the torsion generators of $\Gamma(1)$ are not in $\Gamma(2)$. Use this to deduce that $\Gamma(2)$ is a triangle (∞, ∞, ∞)-group.

(3) Use the mapping properties of the $\wp$-function to show that every complex torus $X(\lambda_1, \lambda_2)$ has a biholomorphic automorphism of order 2 with four fixed points corresponding to the four points of order 2 on the torus, $\wp$-images (denoted $e_0, e_1, \dots$) of the points

$$0, \quad \omega_1 = \lambda_1/2, \quad \omega_2 = \lambda_2/2, \quad \omega_3 = (\lambda_1 + \lambda_2)/2.$$

Normalising so that $w = \lambda_1/\lambda_2$, define Legendre's λ-function as the cross-ratio

$$\lambda(w) = \{e_1, e_2; e_3, e_0\} = \frac{e_3 - e_1}{e_3 - e_2}.$$

Show that under modular transformations $w \mapsto \gamma(w) = \frac{aw+b}{cw+d} \in \Gamma(1)$, the four functions $e_j(\gamma(w))$ are a permutation of the $e_j(w)$. By evaluating the permutation for the generators T and U of $\Gamma(1)$, show that the λ-function is invariant under the congruence subgroup $\Gamma(2)$.

(4) How is the Legendre λ-function related to the j-function? This involves some knowledge of algebraic function fields and Galois theory. (See for instance [19, pp. 293–296].)

(5) Show that the congruence subgroup $\Gamma(2)$ is distinct from the commutator subgroup $H(1) = [\Gamma(1), \Gamma(1)]$ of $\Gamma(1)$, although they have the same index in $\Gamma(1)$.

What can you discover about the quotient surface $\mathcal{U}/H(1)$?

(6) Torus knots. Show that the straight line segment from the origin to the lattice point $p + iq \in L(1, i)$ with g.c.d.$(p.q) = 1$ projects in the quotient torus X_L to a loop winding p times around one generating circle while it winds q times around the transverse one. What happens when g.c.d. $(p, q) \neq 1$?

(7) (For finite group theorists.) The finite simple groups include the important class of *sporadic groups*, with the Fischer–Griess Monster M as largest member. Determine which of them is an epimorphic (i.e., surjective) image of the modular group $\Gamma(1)$. See for instance the book [36] for an interesting geometric account of these groups.

Consider the Riemann surface obtained from the kernel of such a morphism: show that it carries an induced action of the corresponding finite group as automorphism group.

References

[1] M. A. Armstrong, On the fundamental group of an orbit space. *Proc. Cambridge Philos. Soc.* **64**, 299–301 (1968).

[2] L. V. Ahlfors, *Complex Analysis*, 3rd edn. McGraw-Hill, New York (1978).

[3] R. C. Alperin, W. Dicks and J. Porti, The boundary of the Gieseking tree in hyperbolic 3-space, *Topology and Appl.* **93**(3), 219–259 (1999).

[4] V. I. Arnold and A. Avez, *Problèmes Ergodiques de la Mechanique Classique*. Gauthier-Villars (1967). [English translation: *Ergodic Problems of Classical Mechanics*. W. Benjamin, New York (1968)].

[5] A. F. Beardon, *Geometry of Discrete Groups*. Graduate Texts in Mathematics. Springer-Verlag (1983).

[6] R. E. Borcherds, What is moonshine? *Proc. I.C.M.*, Vol. I (Berlin, 1998). Doc. Math., Extra Vol. I, 607–615 (1998).

[7] A. Borel, S. Chowla, C. S. Herz, K. Iwasawa and J.-P. Serre, *Seminar on Complex Multiplication*. Lecture Notes in Mathematics, Vol. 21. Springer-Verlag (1966).

[8] J. M. Borwein, P. B. Borwein and D. H. Bailey, Ramanujan, modular equations and approximations to π. *Amer. Math. Monthly* **96**, 201–219 (1989).

[9] C. Caratheodory, *Geometric Function Theory*, 2 Vols. Chelsea Publ. Co., New York (2nd edn. 1958, 1960). [English translation by F. Steinhardt of original edition, *Functiontheorie*. Birkhäuser (1950)].

[10] J. H. Conway and S. P. Norton, Monstrous moonshine. *Bull. London Math. Soc.* **11**, 308–339 (1979).

[11] H. S. M. Coxeter and W. O. J. Moser, *Generators and Relations for Discrete Groups*, 3rd edn. Ergebnisse der Mathematik und ihrer Grenzgebiete, Band 14. Springer (1972).

[12] H. Davenport, *The Higher Arithmetic*, 4th edn. Hutchinson Press, London (1970).

[13] L. R. Ford, *Automorphic Functions*. Chelsea Books, New York (1951).

[14] I. Frenkel, J. Lepowsky and A. Meurman, *Vertex Operator Algebras and the Monster*. Academic Press (1988).

[15] T. Gannon, Monstrous Moonshine: The First 25 Years. *Bull. London Math. Soc.* **38**, 1–33 (2006).

[16] E. Girondo and G. Gonzalez-Diez, *Introduction to Compact Riemann Surfaces and Dessins*. LMS Student Text, Vol. 79, Cambridge University Press (2012).

[17] A. Grothendieck, Esquisse d'un Programme, In *Geometric Galois Actions*. London Mathematical Society Lecture Notes, Vol. 242, pp. 5–48 (1997).

[18] Y. Hellegouarch, *Invitation to the Mathematics of Fermat–Wiles*. Academic Press (2001).

[19] G. Jones and D. Singerman, *Complex Functions*. Cambridge University Press (1980).

[20] A. W. Knapp, *Elliptic Curves*. Princeton Mathematical Notes, Vol. 40. Princeton University Press (1992).

[21] A. Marden, *Outer Circles: An Introduction to Hyperbolic 3-Manifolds*. Cambridge University Press (2007).
[22] J. W. Milnor, *Introduction to Algebraic K-theory*. Annals of Mathematics Studies, No. 72, Princeton University Press (1971).
[23] J. W. Milnor, On the Brieskorn (p, q, r)-manifolds, In *Knots, Groups and 3-Manifolds*, pp. 175–225. Annals of Mathematics Studies, Vol. 84. Priceton University Press (1975).
[24] Y. N. Minsky, The classification of punctured torus groups. *Ann. of Math.* **149**, 559–626 (1999).
[25] H. Rademacher, *Higher Mathematics from an Elementary Point of View*. Birkhäuser (1983).
[26] E. G. Rees, *Lectures on Geometry*. Universitext Series, Springer (1983; 6th printing 2013).
[27] R. E. Schwartz, Pappus's theorem and the modular group. *Publ. Math. Inst. Hautes Études Sci.* **78**, 187–206 (1993).
[28] J.-P. Serre, *A Course in Arithmetic*. Graduate Text in Mathematics. Springer (1973).
[29] J.-P. Serre, *Trees*. Springer (1981). [English translation of Arbres, amalgames et SL(2). *Astérisque* (1973)].
[30] G. B. Shabat and V. I. Voevodsky, Drawing curves over number fields. In *The Grothendieck Festschrift*, Vol. III, pp. 299–327. Progress in Mathematics, Vol. 88. Birkhäuser (1990).
[31] G. Shimura, *Arithmetic Theory of Automorphic Functions*. Princeton University Press and Iwanami-Shoten (1971).
[32] C. L. Siegel, *Topics in Complex Function Theory*, 3 Vols. Wiley-Interscience, New York (1969, 1973).
[33] J. H. Silverman, *Advanced Topics in the Arithmetic of Elliptic Curves*. Graduate Texts in Mathematics, Vol. 151. Springer-Verlag (1994).
[34] W. P. Thurston, *Three-Dimensional Geometry and Topology*. Princeton Mathematical Series, Vol. 35. Princeton University Press (1997).
[35] A. Wiles, Modular elliptic curves and Fermat's last theorem. *Ann. of Math.* **141**, 443–551 (1995).
[36] R. A. Wilson, *The Finite Simple Groups*. Graduate Texts in Mathematics, Vol. 251. Springer-Verlag (2009).
[37] D. Zagier, Elliptic modular forms & their applications, Chapter 1. In *The 1-2-3 of Modular Forms*. Universitext. Springer (2008).
[38] A. Verjovsky, *Arithmetic, Geometry and dynamics in the modular orbifold*, in Dynamical Systems, (Santiago de Chile 1990), Pitman Research Note Series. See also preprint from ICTP, Trieste, (July, 1991) IC/91/93.

Chapter 3

Geometric Group Theory

Ian Chiswell

School of Mathematical Sciences
Queen Mary University of London, London E1 4NS, UK
i.m.chiswell@qmul.ac.uk

This is an introduction to the theory of group presentations and related constructions (free groups, free products and HNN-extensions). Connections with topology via CW-complexes are briefly explored, and there is an account of the Bass–Serre theory of group actions on trees. Some of the interesting classes of groups arising from the topological connections are described.

1. Free Groups and Presentations

The study of presentations of groups by generators and relations has for some time been called combinatorial group theory. The subject has many interactions with topology and geometry, because an action of a group on a space can give rise to a presentation of the group, and the fundamental group of a space can in some cases be described by a presentation. Conversely, a presentation of a group can give rise to an action of the group on a space. The study of these interactions is now called geometric group theory. In this first section, the idea of a presentation is introduced.

Let X be a subset of a group G. The subgroup generated by X, denoted by $\langle X \rangle$, is the intersection of all subgroups of G containing X as a subset.

If $g \in G$, then $g \in \langle X \rangle$ if and only if can be written as $g = x_1^{e_1} \ldots x_n^{e_n}$ for some $x_i \in X$, $e_i = \pm 1$ and $n \geq 0$. (When $n = 0$, this means $g = 1$.) This is because the set of all such products is a subgroup of G, and any subgroup of G containing X contains all such products, by induction on n.

Definition. *X generates G if $G = \langle X \rangle$.*

If $f : G \to H$ is a homomorphism and $G = \langle X \rangle$, then
$$f(x_1^{e_1} \ldots x_n^{e_n}) = f(x_1)^{e_1} \ldots f(x_n)^{e_n}$$
so f is determined by its restriction to X.

If $G = \langle X \rangle$ and $f : X \longrightarrow H$ is a map to a group H, then f does not necessarily extend to a homomorphism $\tilde{f} : G \to H$. For example, let $X = \{x\}$, let $G = \langle x \rangle$ be cyclic of order m and $H = \langle y \rangle$ cyclic of order n, where m and n are relatively prime and greater than 1. Define $f : X \longrightarrow H$ by $f(x) = y$, and suppose an extension $\tilde{f}$ exists. There are integers a, b such that $am + bn = 1$. Then $x = x^{bn}$, so $y = f(x) = f(x)^{bn} = y^{bn} = 1$, a contradiction.

In general, for an extension to exist, the following condition is needed:
$$x_1^{e_1} \ldots x_n^{e_n} = 1 \ (x_i \in X, \ e_i = \pm 1) \Rightarrow f(x_1)^{e_1} \ldots f(x_n)^{e_n} = 1.$$
Then it is easy to verify that there is a well-defined mapping $\tilde{f} : G \longrightarrow H$, given by $\tilde{f}(x_1^{e_1} \ldots x_n^{e_n}) = f(x_1)^{e_1} \ldots f(x_n)^{e_n}$, which is a homomorphism.

Free groups

Definition. Let X be a subset of a group F. Then F is a *free group with basis X* if, for any map $f : X \to H$, where H is a group, there is a unique extension of f to a homomorphism $\tilde{f} : F \to H$.

Note. It follows that X generates F. For let $F_1 = \langle X \rangle$. The inclusion map $X \to F_1$ has an extension to a homomorphism $f_1 : F \to F_1$. Let $f_2 : F_1 \to F$ be the inclusion map. Then $f_2 f_1$ and id_F both extend the inclusion map $X \to F$, so $f_2 f_1 = \mathrm{id}_F$, hence f_2 is onto, i.e., $F = F_1$.

Proposition 1.1. *Let F_1, F_2 be free groups with bases X_1, X_2. Then F_1 is isomorphic to F_2 if and only if $|X_1| = |X_2|$. ($|X_i|$ means the cardinality of X_i.)*

Proof. Suppose $|X_1| = |X_2|$ and let $f : X_1 \to X_2$ be a bijective map. Viewed as a mapping $X_1 \to F_2$, there is an extension of f to a homomorphism $\tilde{f} : F_1 \to F_2$. Similarly, putting $g = f^{-1}$, g has an extension to a homomorphism $\tilde{g} : F_2 \to F_1$. Both $\tilde{g}\tilde{f}$ and $\mathrm{id}_{F_1} : F_1 \to F_1$ extend the inclusion map $X_1 \to F_1$, so $\tilde{g}\tilde{f} = \mathrm{id}_{F_1}$. Similarly, $\tilde{f}\tilde{g} = \mathrm{id}_{F_2}$, hence $\tilde{f}$ is an isomorphism.

For the converse, let F be free with basis X, let F^2 be the subgroup generated by $\{u^2 \mid u \in F\}$, which is a normal subgroup of F, and let V be a vector space over $\mathbb{Z}/2\mathbb{Z}$ with basis X. The inclusion mapping $X \to V$ extends to a homomorphism $F \to V$ of groups; the kernel contains F^2, so there is an induced homomorphism $F/F^2 \to V$, which is a linear map of vector spaces. It follows that the projection homomorphism $F \to F/F^2$ is injective on X and the image of X is a $\mathbb{Z}/2\mathbb{Z}$-basis for F/F^2. Hence $|X| = \dim_{\mathbb{Z}/2\mathbb{Z}}(F/F^2)$.

Now an isomorphism from F_1 to F_2 induces an isomorphism from F_1/F_1^2 to F_2/F_2^2 (of vector spaces), so $|X_1| = |X_2|$. $\square$

Definition. If F is free with basis X, then $|X|$ is the *rank* of F.

For example, the trivial group is free of rank 0 (with basis the empty set), and the infinite cyclic group is free of rank 1 (with basis any of the two generators).

Existence. Let X be a set. Let X^{-1} be a set in one-to-one correspondence with X via a map $x \mapsto x^{-1}$, and with $X \cap X^{-1} = \emptyset$. Put $X^{\pm 1} = X \cup X^{-1}$ and define $(x^{-1})^{-1} = x$, for $x \in X$, to obtain an involution $X^{\pm 1} \to X^{\pm 1}$ without fixed points.

A *word* in $X^{\pm 1}$ is a finite sequence $w = (a_1, \ldots, a_n)$, where $a_i \in X^{\pm 1}$ for $1 \leq i \leq n$, and $n \geq 0$. (When $n = 0$, $w = 1$, the *empty word*.) The *length* of w, denoted $|w|$, is n (note that $|1| = 0$).

Let W be the set of words. If $w' = (b_1, \ldots, b_m)$ is also in W, define

$$w.w' = (a_1, \ldots, a_n, b_1, \ldots, b_m)$$

($1.w = w.1 = w$), giving an associative binary operation on W, with identity element 1. Define $w^{-1} = (a_n^{-1}, \ldots, a_1^{-1})$ ($1^{-1} = 1$). Then $(u.v)^{-1} = v^{-1}.u^{-1}$ for all $u, v \in W$.

For $u, v \in W$, define $u \simeq v$ to mean v is obtained from u by inserting or deleting a part aa^{-1}, where $a \in X^{\pm 1}$ (e.g., $xyzz^{-1}x \simeq xyx \simeq xz^{-1}zyx$).

For $u, v \in W$, define $u \sim v$ to mean that there is a sequence $u = u_1, u_2, \ldots, u_k = v$, where $u_i \simeq u_{i+1}$ for $1 \leq i \leq k-1$. This is an equivalence relation on W.

If $u \simeq v$ and $w \in W$, then $u.w \simeq v.w$, hence if $u \sim v$ then $u.w \sim v.w$. Similarly, $u \sim v$ implies $w.u \sim w.v$. Hence, if $u \sim u_1$ and $v \sim v_1$, then $u.v \sim u.v_1$ and $u.v_1 \sim u_1.v_1$, so $u.v \sim u_1.v_1$.

If $u \simeq v$ then $u^{-1} \simeq v^{-1}$, hence if $u \sim v$ then $u^{-1} \sim v^{-1}$. Also, $u.u^{-1} \sim 1 \sim u^{-1}.u$.

Let $[u]$ denote the equivalence class of $u \in W$ and let F denote the set of equivalence classes. It follows that F is a group, by defining $[u][v] = [u.v]$ with identity element $[1]$, and $[u]^{-1} = [u^{-1}]$.

Definition. A word $w \in W$ is *reduced* if it contains no part aa^{-1} with $a \in X^{\pm 1}$.

Theorem 1.2 (Normal Form Theorem). *Every equivalence class contains a unique reduced word.*

Proof. An equivalence class contains a word of minimal length, which must be reduced, and it remains to show uniqueness.

Suppose $[u] = [v]$, where u, v are reduced, so there is a sequence $u = u_1, \ldots, u_k = v$ with $u_i \simeq u_{i+1}$. Choose such a sequence with $n := \sum_{i=1}^{k} |u_i|$ as small as possible. It will be shown that $k = 1$, so $u = v$, as required.

Suppose $k \geq 2$ and let u_i be a word of maximal length in the sequence. Then $1 < i < k$ since u, v are reduced. Moreover, u_i is obtained from u_{i-1} by inserting yy^{-1} for some $y \in X^{\pm 1}$, and u_{i+1} is obtained from u_i by deleting zz^{-1} for some $z \in X^{\pm 1}$. If the parts yy^{-1} and zz^{-1} of u_i coincide or overlap by a single letter, then $u_{i-1} = u_{i+1}$, and u_i, u_{i+1} can be omitted from the sequence, contradicting the minimality of n.

Otherwise, u_i can be replaced in the sequence by u'_i, where u'_i is obtained from u_i by deleting zz^{-1}, and u_{i+1} is obtained from u'_i by inserting yy^{-1}. This also contradicts the minimality of n, and completes the proof. □

Consequently, the map $X \to F$, $x \mapsto [x]$ is injective, as (x) is a reduced word, and x can be identified with $[x]$, for $x \in X$.

Theorem 1.3. *The group F is free with basis X.*

Proof. Let $f : X \to H$ be a map, and let H be a group. Extend f to $\bar{f} : W \to H$ by:

$$\bar{f}(x_1^{e_1}, \ldots, x_n^{e_n}) = f(x_1)^{e_1} \ldots f(x_n)^{e_n}$$

($x_i \in X$, $e_i = \pm 1$) and $\bar{f}(1) = 1$. If $u \simeq v$ then $\bar{f}(u) = \bar{f}(v)$, so if $u \sim v$ then $\bar{f}(u) = \bar{f}(v)$. So a map $\tilde{f} : F \to H$ can be defined by $\tilde{f}([u]) = \bar{f}(u)$. If $u, v \in W$ then $\bar{f}(u.v) = \bar{f}(u)\bar{f}(v)$, hence $\tilde{f}$ is a homomorphism extending f.

Finally, X generates F, so the extension of f is unique. □

Presentations

Let X be a set, W the set of words in $X^{\pm 1}$, G a group, and $\alpha : X \to G$ a map. Extend α to $\bar{\alpha} : W \to G$ by: $\bar{\alpha}(x_1^{e_1}, \ldots, x_n^{e_n}) = \alpha(x_1)^{e_1} \ldots \alpha(x_n)^{e_n}$.

Definition. Let R be a subset of W. The group G has presentation $\langle X \mid R \rangle$ (via α) if

(1) $\bar{\alpha}(w) = 1$ for all $w \in R$;
(2) given a group H and map $f : X \to H$, such that $\bar{f}(w) = 1$ for all $w \in R$, there is a unique homomorphism $\varphi : G \to H$ such that $\varphi\alpha = f$.

The requirement that φ is unique is equivalent to the statement that $\alpha(X)$ generates G. If $\alpha(X)$ generates G, then clearly φ is unique, and the converse follows by an argument similar to that after the definition of free group.

For any X and $R \subseteq W$, there is a group with presentation $\langle X \mid R \rangle$. Let F be free with basis X; then R represents a subset of F. Let N be the normal subgroup of F generated by R (i.e., the subgroup generated by all conjugates of elements of R), let $G = F/N$, and let $\alpha : X \to G$ be the restriction to X of the projection map $F \to F/N$. Then $\langle X \mid R \rangle$ is a presentation of G via α.

The map α often suppressed to simplify notation, but care is needed since it need not be injective (e.g., R might contain $x.y^{-1}$, where $x, y \in X$, $y \neq x$). If α is an inclusion map, then φ is an extension of f to G.

The elements of R are called *relators* of the presentation. If $w \in R$, one often writes $w = 1$ instead of just w. More generally, if $w = w_1.w_2^{-1}$, one can write $w_1 = w_2$ instead of w. This is called a *relation*. If $X = \{x_1, \ldots, x_m\}$, and $R = \{w_1, \ldots, w_n\}$, the presentation is written as $\langle x_1, \ldots, x_m \mid w_1, \ldots, w_n \rangle$, omitting the braces.

Note that every group has a presentation. Let X be a set of generators for G, let F be free with basis X, and let $f : F \to G$ be the extension of the inclusion map $X \to G$ to F. Let $N = \ker(f)$, viewed as a set of reduced words. Then $\langle X \mid N \rangle$ is a presentation of G via the inclusion map $X \to G$.

Also, if G, H both have presentation $\langle X \mid R \rangle$ via suitable maps, then $G \cong H$, and if $G \cong H$, then any presentation for G is one for H.

Examples. (1) If G has presentation $\langle x \mid x^n \rangle$, then G is cyclic of order n. For G is clearly cyclic of order at most n, and if $\langle a \rangle$ is cyclic of order n, the map $x \mapsto a$ extends to a homomorphism $G \longrightarrow \langle a \rangle$ which is onto, so $|G| \geq n$. In what follows, C_n will denote a cyclic group of order n.

(2) Let $n \geq 2$ and let a be a rotation of the plane $\mathbb{R}^2 = \mathbb{C}$ anticlockwise through $2\pi/n$ ($z \mapsto ze^{2\pi i/n}$). Let b be reflection in the real axis ($z \mapsto \bar{z}$), so a has order n, b has order 2, and $bab^{-1} = a^{-1}$. Define D_n to be $\langle a, b \rangle$ (the subgroup generated by a, b in the group of isometries of $\mathbb{R}^2$). Let

$A = \langle a \rangle$, $B = \langle b \rangle$; then $A \trianglelefteq D_n$, so $D_n = AB$, and $A \cap B = \{1\}$. Hence $|D_n| = |A||B| = 2n$.

The group D_n, sometimes written D_{2n}, is called the *dihedral group* of order $2n$. It is the set of all isometries of $\mathbb{R}^2$ which map the set $\{e^{2\pi ik/n}\}$ of complex nth roots of 1 onto itself, equivalently the regular polygon with these points as vertices if $n \geq 3$.

In fact, $\langle x, y \mid x^n, y^2, yxy^{-1} = x^{-1} \rangle$ is a presentation of D_n, via $\alpha : x \mapsto a$, $y \mapsto b$. Let G be the group with this presentation; then α extends to a homomorphism $G \xrightarrow{\text{onto}} D_n$, and it is enough to show that $|G| \leq 2n$, so that α is an isomorphism. Let H be the subgroup of G generated by x; since $x^n = 1$ in G, $|H| \leq n$. Now $Hx = H$, $Hyx = Hx^{-1}y = Hy$, $Hy = Hy$, $(Hy)y = Hy^2 = H$, and G is generated by $\{x, y\}$, so any element of G permutes $\{H, Hy\}$ by right multiplication. This action on the cosets of H is transitive, so $\{H, Hy\}$ is a complete list of the cosets. Hence $(G : H) \leq 2$, so $|G| = |H|(G : H) \leq 2n$.

(3) The symmetric group S_n has the presentation

$$\langle x_1, \ldots, x_{n-1} \mid x_i^2 = 1 \ (i \leq n-1), (x_i x_{i+1})^3 = 1 \ (i \leq n-2),$$
$$(x_i x_j)^2 = 1 \ (j < i - 1)\rangle$$

via the map sending x_i to the transposition $(i, i+1)$. See [4, Exercise 3, Chapter 5].

(4) $\langle X \mid \emptyset \rangle$ is a presentation of the free group on X.

(5) Let G be any group, and let $X = \{x_g \mid g \in G\}$ be a set in one-to-one correspondence with G via $g \mapsto x_g$. Let $R = \{x_g.x_h.x_{gh}^{-1} \mid g, h \in G\}$. Let H have presentation $\langle X \mid R \rangle$. There is a homomorphism $H \xrightarrow{\text{onto}} G$ induced by $x_g \mapsto g$ ($g \in G$), and this is an isomorphism. (To see this, first show that, in H, $x_{g_1}^{e_1} \ldots x_{g_n}^{e_n} = x_{g_1^{e_1} \ldots g_n^{e_n}}$.) Hence $\langle X \mid R \rangle$ is a presentation of G, called the *standard presentation* of G, denoted by $\langle G \mid \text{rel}(G) \rangle$.

Tietze transformations

Let $\langle X \mid R \rangle$ be a presentation of G via α. Let F be free on X, and N be the normal subgroup of F generated by R. Words in $X^{\pm 1}$ representing elements of N are called *consequences* of R. Equivalently, w is a consequence of R if and only if $\bar{\alpha}(w) = 1$.

This can be applied to relations: $w_1 = w_2$ is a consequence of R if $w_1.w_2^{-1}$ is a consequence of R, equivalently $\bar{\alpha}(w_1) = \bar{\alpha}(w_2)$. Similarly, a relation or relator is a consequence of a set of relations (or a mixture of relations and relators) if it is a consequence of the corresponding set of relators.

Definition. A *Tietze transformation* of $\langle X \mid R \rangle$ is one of the following.

(T1) Replace $\langle X \mid R \rangle$ by $\langle X \cup Y \mid R \cup \{y = w_y \mid y \in Y\} \rangle$, where Y is a set with $X \cap Y = \emptyset$ and for each $y \in Y$, w_y is a word in $X^{\pm 1}$.
(T2) The inverse of T1.
(T3) Replace $\langle X \mid R \rangle$ by $\langle X \mid R \cup S \rangle$, where S is a set of words in $X^{\pm 1}$ which are consequences of R.
(T4) The inverse of T3.

Theorem 1.4. *Two presentations $\langle X \mid R \rangle$ and $\langle Y \mid S \rangle$ are presentations of the same group if and only if one can be obtained from the other by a finite succession of Tietze transformations.*

Proof. See [5, Theorem 15, Chapter 1] or [12, Proposition 2.1, Chapter II]. □

Example. The presentation $\langle x, y \mid yxy^{-1}x,\ x^n, y^2 \rangle$ of D_n can be transformed as follows:

$$\xrightarrow{T1} \langle x, y, u \mid yxy^{-1}x,\ x^n,\ y^2,\ u = yx \rangle$$
$$\xrightarrow{T3} \langle x, y, u \mid yxy^{-1}x,\ x^n,\ y^2,\ u = yx,\ u^2 = 1 \rangle$$
$$\xrightarrow{T4} \langle x, y, u \mid x^n,\ y^2,\ u = yx,\ u^2 = 1 \rangle$$
$$\xrightarrow{T3 \& T4} \langle x, y, u \mid x^n,\ y^2,\ u^2,\ x = y^{-1}u \rangle$$
$$\xrightarrow{T3 \& T4} \langle x, y, u \mid (y^{-1}u)^n,\ y^2,\ u^2,\ x = y^{-1}u \rangle$$
$$\xrightarrow{T2} \langle y, u \mid (y^{-1}u)^n,\ y^2,\ u^2 \rangle$$
$$\xrightarrow{T3 \& T4} \langle y, u \mid y^2,\ u^2,\ (yu)^n \rangle$$
$$\longrightarrow \langle a, b \mid a^2,\ b^2,\ (ab)^n \rangle.$$

Exercises

1.1 Let X be a subset of a group G. Prove that G is free with basis X if and only if: X generates G, and no non-empty reduced word in $X^{\pm 1}$ represents the identity element of G. [The element of G represented by a word $(a_1, \ldots, a_n)$, where $a_i \in X^{\pm 1}$, is $a_1 \ldots a_n$ (product in G).]

1.2 Let F be free with basis $\{x, y\}$. Show that the set $Y = \{x^i y x^{-i} \mid i \in \mathbb{N}\}$ is a basis for the subgroup of F it generates. (*Hint*: use Exercise 1.1.) Thus a free group of rank 2 contains a free group of countably infinite rank.

1.3 Show that the following are presentations of the trivial group.
   (a) $\langle x, y \mid xyx^{-1} = y^2, yxy^{-1} = x^2 \rangle$;
   (b) $\langle x, y, z \mid xyx^{-1} = y^2, yzy^{-1} = z^2, zxz^{-1} = x^2 \rangle$.

1.4 Transform the first presentation to the second by Tietze transformations.
   (a) $\langle a, b \mid aba = bab \rangle$ to $\langle x, y \mid x^2 = y^3 \rangle$;
   (b) $\langle a, b, c, d \mid ab = c, bc = d, cd = a, da = b \rangle$ to $\langle a \mid a^5 = 1 \rangle$.

2. Two Basic Constructions

The two constructions in question are very useful in dealing with presentations. The account here follows [4, pp. 99–104]. The first arises from the following problem. Given a family of groups, each of which has a subgroup isomorphic to a given group A, embed the family in a group in which these isomorphic copies of A (and no other elements) are identified. It will be shown that the *free product with amalgamation*, defined below, has the required property.

Free products with amalgamation. Let $\{G_i \mid i \in I\}$ be a family of groups, A a group and $\alpha_i : A \longrightarrow G_i$ a monomorphism, for all $i \in I$. A group G is the *free product of the G_i with A amalgamated* (via the α_i) if there are homomorphisms $f_i : G_i \longrightarrow G$ such that $f_i \alpha_i = f_j \alpha_j$ for all $i, j \in I$, and if $h_i : G_i \longrightarrow H$ are homomorphisms with $h_i \alpha_i = h_j \alpha_j$ for all $i, j \in I$, then there is a unique homomorphism $h : G \longrightarrow H$ such that $h f_i = h_i$ for all $i \in I$.

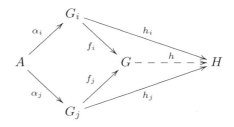

Such a group G is unique up to isomorphism by a simple argument from elementary category theory, and a similar argument shows that G is generated by $\bigcup_{i \in I} f_i(G_i)$. For details, see [4, p. 100].

The existence of such a group G can be demonstrated using presentations. Let $\langle X_i \mid R_i \rangle$ be a presentation of G_i, with $X_i \cap X_j = \emptyset$ for $i \neq j$. Let Y be a set of generators for A and, for each $y \in Y$, $i \in I$, let $a_{i,y}$ be a word in $X_i^{\pm 1}$ representing $\alpha_i(y)$. Let $S = \{a_{i,y}\, a_{j,y}^{-1} \mid y \in Y,\ i,j \in I,\ i \neq j\}$ and let G be the group with presentation $\langle \bigcup_{i \in I} X_i \mid \bigcup_{i \in I} R_i \cup S \rangle$, via φ, say. Then G is the free product with amalgamation. For the restriction $\varphi|_{X_i} : X_i \longrightarrow G$ induces a homomorphism $f_i : G_i \longrightarrow G$ (by definition of a presentation) with $f_i \alpha_i = f_j \alpha_j$ for $i, j \in I$. The existence and uniqueness of the mapping h in the definition again follows by definition of a presentation.

If G is the free product of $\{G_i \mid i \in I\}$ with A amalgamated, this is expressed by writing $G = \bigstar_A G_i$, or in the case of two groups $\{G_1, G_2\}$, $G = G_1 *_A G_2$.

Example. Let $G_1 = \langle x \rangle$, $G_2 = \langle y \rangle$, $A = \langle a \rangle$ be infinite cyclic groups, and let α_1, α_2 be the homomorphisms defined by $\alpha_1 : a \mapsto x^m$, $\alpha_2 : a \mapsto y^n$, where $m, n > 1$. Then, after Tietze transformations, $G = G_1 *_A G_2$ has presentation $\langle x, y \mid x^m = y^n \rangle$. In the case $m = 2$, $n = 3$, the presentation transforms by Tietze transformations to $\langle a, b \mid aba = bab \rangle$, and G is a group discussed in Section 5, the braid group on three strings.

Structure of free products with amalgamation. Let $G = \bigstar_A G_i$ via $\alpha_i : A \to G_i$, $i \in I$. To represent elements of G by certain words, identify A with $\alpha_i(A)$ and assume $G_i \cap G_j = A$ for $i \neq j$.

Define a *word* to be a finite sequence $w = (g_1, \ldots, g_k)$, where $k \geq 1$ and $g_1 \in G_{i_1}, \ldots, g_k \in G_{i_k}$ for some $i_1, \ldots, i_k \in I$. The element of G *represented* by w is $f_{i_1}(g_1) \ldots f_{i_k}(g_k)$. Since $\bigcup_{i \in I} f_i(G_i)$ generates G and the f_i are homomorphisms, every element of G is represented by some word.

Definition. Let $w = (g_1, \ldots, g_k)$ be a word with $g_j \in G_{i_j}$, $1 \leq j \leq k$. Then w is *reduced* if no consecutive pair g_i, g_{i+1} belongs to the same group G_i, that is, $i_j \neq i_{j+1}$ for $1 \leq j \leq k-1$ and $g_j \notin A$ for $1 \leq j \leq k$, unless $k = 1$.

If w is not reduced, one can replace w by $(g_1, \ldots, g_j g_{j+1}, \ldots, g_k)$, for some j, a word representing the same element of G. Hence every element of G is represented by a reduced word (take a word of shortest length representing it). Unfortunately, the representation is not necessarily unique;

for example, if (g_1, g_2) is reduced of length 2 and $a \in A$, $a \neq 1$, then $(g_1 a, a^{-1} g_2)$ is reduced and represents the same element of G. To obtain uniqueness, it is necessary to choose coset representatives.

For every $i \in I$, choose a transversal T_i for the cosets $\{Ax \mid x \in G_i\}$, with $1 \in T_i$.

Definition. A *normal* word is a word $(a, r_1, \ldots, r_k)$ where $a \in A$, $k \geq 0$, $r_j \in T_{i_j} \setminus \{1\}$ for some $i_j \in I$ $(1 \leq j \leq k)$, and $i_j \neq i_{j+1}$ for $1 \leq j \leq k - 1$.

Theorem 2.1 (Normal Form Theorem). *Any element of G is represented by a unique normal word.*

Proof. Let $g \in G$ be represented by the reduced word $w = (g_1, \ldots, g_k)$ with $g_j \in G_{i_j}$. One can write

$$g_k = a_k r_k, \ g_{k-1} a_k = a_{k-1} r_{k-1}, \ldots, \ g_1 a_2 = a_1 r_1,$$

where $a_j \in A$ and $r_j \in T_{i_j}$ for $1 \leq j \leq k$. Then $(a_1, r_1, r_2, \ldots, r_k)$ is a normal word representing g. (Note that $a \in A$ is represented by the normal word (a), in particular (1) represents the identity element of G.)

It remains to show uniqueness, and for this see [4, Theorem 5.8]. The idea (due to van der Waerden) is to define an action of G on the set W of normal words, equivalently a homomorphism $h : G \longrightarrow S(W)$, the symmetric group on W. By the defining property of G, it suffices to define, for $i \in I$, a homomorphism $h_i : G_i \longrightarrow S(W)$ with $h_i \alpha_i = h_j \alpha_j$ for all $i, j \in I$. This is done by considering the effect on the normal form when an element of G is left-multiplied by an element of G_i, but using the elegant method in Serre's tree notes [17, Chapter I, Theorem 1]. The proof concludes by noting that, if $g \in G$ is represented by the normal word $w = (a, r_1, \ldots, r_k)$, then $h(g)(1) = w$, so w is uniquely determined by g. □

This has two simple consequences.

Corollary 2.2.

(1) *The homomorphisms f_i are injective.*
(2) *No reduced word of length greater than 1 represents the identity element of G.*

By Corollary 2.2(1), the homomorphisms f_i can be suppressed.

Free products: An important special case is the case $A = \{1\}$, when G is called the *free product* of the family $\{G_i \mid i \in I\}$, written $G = \bigast_{i \in I} G_i$ (or $G = G_1 * G_2$ for two groups). The definition simplifies to: given any collection of homomorphisms $h_i : G_i \longrightarrow H$, there is a unique extension to a homomorphism $h : G \longrightarrow H$. Let $\langle X_i \mid R_i \rangle$ be a presentation of G_i, with $X_i \cap X_j = \emptyset$ for $i \neq j$. Then $\mathop{\mathlarger{\ast}}_{i \in I} G_i$ has the presentation $\langle \bigcup_{i \in I} X_i \mid \bigcup_{i \in I} R_i \rangle$ (obtained from the general presentation of a free product with amalgamation by taking the presentation of the trivial group with empty set of generators and empty set of relators.)

A reduced word is a word $(g_1, \ldots, g_k)$ such that $g_j \in G_{i_j}$, $g_j \neq 1$ unless $k = 1$, and $i_j \neq i_{j+1}$ for $1 \leq j \leq k - 1$. The Normal Form Theorem can be replaced by the simpler statement that every element of G is represented by a unique reduced word. Also, the G_i embed in G.

Examples. (1) $\langle x, y \mid x^r = 1, \ y^s = 1 \rangle$ is a presentation of $C_r \ast C_s$. A special case is $r = s = 2$, the *infinite dihedral group*, which has the dihedral group D_n as a homomorphic image, for all $n \geq 2$ (using the presentation of D_n in the example after Theorem 1.4). Another is $r = 2, s = 3$; this group is called the *modular group*, and is isomorphic to $\mathrm{PSL}_2(\mathbb{Z})$ (see [13, Theorem 3.1]).

(3) Let F be a free group with basis $\{x_i \mid i \in I\}$; then $F = \mathop{\mathlarger{\ast}}_{i \in I} \langle x_i \rangle$, a free product of infinite cyclic groups.

HNN-extensions. Suppose B, C are subgroups of a group A and $\gamma : B \longrightarrow C$ is an isomorphism. The second basic construction embeds A in a group G so that γ is the restriction of an inner automorphism of G.

Let G have presentation $\langle \{t\} \cup X \mid R_1 \cup R_2 \rangle$, where $X = \{x_g \mid g \in A\}$, $t \notin X$, $R_1 = \{x_g x_h x_{gh}^{-1} \mid g, h \in A\}$ and $R_2 = \{tx_b t^{-1} x_{\gamma(b)}^{-1} \mid b \in B\}$.

There is a homomorphism $f : A \to G$ given by $a \mapsto x_a$; it will be shown that f is injective.

Definition. The group G is called a Higman–Neumann–Neumann extension (abbreviated to HNN-extension) with *base* A, *associated pair* of subgroups B, C and *stable letter* t.

The presentation is often abbreviated to $\langle t, A \mid \mathrm{rel}(A), tBt^{-1} = \gamma(B) \rangle$. More generally, let $\langle Y \mid S \rangle$ be a presentation of A, let $\{b_j \mid j \in J\}$ be a set of words in $Y^{\pm 1}$ representing a set of generators for B, and let c_j be a word representing $\gamma(b_j)$ (more accurately, γ(the generator represented by b_j)). Then $\langle \{t\} \cup Y \mid S \cup \{tb_j t^{-1} = c_j \mid j \in J\} \rangle$ is also a presentation of G (exercise).

Example. The *Baumslag–Solitar group* $\mathrm{BS}(r, s)$ is the group with presentation $\langle x, y \mid xy^r x^{-1} = y^s \rangle$. It is an HNN-extension, with base an infinite cyclic group $A = \langle y \rangle$, stable letter x and associated pair $\langle y^r \rangle$, $\langle y^s \rangle$. These groups, introduced in [1], have some interesting properties and have been studied by many authors.

Let G be an HNN-extension as above. To describe the structure of G, define a *word* to be a finite sequence $(g_0, t^{e_1}, g_1, t^{e_2}, g_2, \ldots, t^{e_k}, g_k)$ where $k \geq 0$, $e_i = \pm 1$ and $g_i \in A$ for $0 \leq i \leq k$. Since G is generated by $f(A) \cup \{t\}$, every element of G is represented in an obvious way by such a word.

Definition. Such a word is *reduced* if it has no part of the form t, b, t^{-1} with $b \in B$ or t^{-1}, c, t with $c \in C$.

If a word is not reduced, it can be replaced by a shorter word representing the same element of G, hence any element of G is represented by a reduced word. However, the representation need not be unique, and once again it is necessary to choose coset representatives to obtain uniqueness.

Choose a transversal T_B for the cosets $\{Bg \mid g \in A\}$ and a transversal T_C for the cosets $\{Cg \mid g \in A\}$, with $1 \in T_B, T_C$.

Definition. A *normal word* is a reduced word

$$(g_0, t^{e_1}, r_1, t^{e_2}, r_2, \ldots, t^{e_k}, r_k),$$

where $g_0 \in A$, $r_i \in T_B$ if $e_i = 1$ and $r_i \in T_C$ if $e_i = -1$.

Theorem 2.3 (Normal Form Theorem). *Every element of G is represented by a unique normal word.*

Proof. Suppose $(g_0, t^{e_1}, g_1, t^{e_2}, g_2, \ldots, t^{e_k}, g_k)$ is a reduced word with $k \geq 1$. If $e_k = 1$, write $g_k = br$ with $b \in B$, $r \in T_B$. Then

$$(g_0, t^{e_1}, g_1, t^{e_2}, g_2, \ldots, g'_{k-1}, t^{e_k}, r),$$

where $g'_{k-1} = g_{k-1}\gamma(b)$, represents the same element of G.

If $e_k = -1$, write $g_k = cr$ with $c \in C$ and $r \in T_C$. Then

$$(g_0, t^{e_1}, g_1, t^{e_2}, g_2, \ldots, g'_{k-1}, t^{e_k}, r),$$

where $g'_{k-1} = g_{k-1}\gamma^{-1}(c)$, represents the same element of G. If $k > 1$, then g'_{k-1} can be written as a similar product (with two cases, depending on the value of e_{k-1}), and continuing in this way leads to a normal word representing the same element of G, so every element of G is represented by a normal word. Uniqueness can be proved by a variant of the van der Waerden method used for free products with amalgamation. See [5, Theorem 31, Chapter 1] or [12, Theorem 2.1, Chapter IV] for details. □

Corollary 2.4.

(1) *The homomorphism $f : A \longrightarrow G$ is injective.*
(2) *(Britton's Lemma) No reduced word $(g_0, t^{e_1}, g_1, t^{e_2}, g_2, \ldots, t^{e_k}, g_k)$ with $k > 0$ represents the identity element of G.* □

More generally, given A and a family $\gamma_i : B_i \longrightarrow C_i$ ($i \in I$) of isomorphisms, where B_i, C_i are subgroups of A, one can form the HNN-extension with presentation (in abbreviated form)

$$\langle t_i \ (i \in I), \ A \mid \mathrm{rel}(A), \ t_i B_i t_i^{-1} = \gamma_i(B_i) \ (i \in I) \rangle.$$

There are generalisations of the Normal Form Theorem and its corollary.

An application

Theorem 2.5. *Any countable group can be embedded in a 2-generator group.*

Proof. Let $G = \{g_0, g_1, g_2, \ldots\}$, where $g_0 = 1$. Let F be free with basis $\{a, b\}$. Then $\langle b^n a b^{-n} \ (n \geq 0) \rangle$, that is, the subgroup generated by the set $\{b^n a b^{-n} \mid n \geq 0\}$, is a free subgroup of F with the given elements as basis (see Exercise 1.2). Similarly $\langle a^n b a^{-n} \ (n \geq 0) \rangle$ is a free subgroup.

In $G * F$, $C := \langle g_n a^n b a^{-n} \ (n \geq 0) \rangle$ is free with basis the given elements. For the homomorphisms $G \to F$, $g \mapsto 1$ ($g \in G$) and $\mathrm{id}_F : F \to F$ have an extension to a homomorphism $f : G * F \to F$. If w is represented by a non-empty reduced word in $\{g_n a^n b a^{-n} \mid (n \geq 0)\}^{\pm 1}$ of positive length, then $f(w)$ is represented by a corresponding reduced word in $\{a^n b a^{-n} \mid (n \geq 0)\}^{\pm 1}$, so $f(w) \neq 1$, hence $w \neq 1$. It follows that C is free as claimed (see Exercise 1.1(b)). Therefore one can form the HNN-extension

$$H := \langle t, G * F \mid \mathrm{rel}(G * F), t(b^n a b^{-n}) t^{-1} = g_n a^n b a^{-n} \ (n \geq 0) \rangle.$$

Then $G \leq G * F \leq H$, $g_n \in \langle t, a, b \rangle$, and (taking $n = 0$) $tat^{-1} = b$, so H is generated by t and a. □

Exercises

2.1 Let $G = A *_C B$, suppose $a \in A$, $b \in B$ and $\langle a \rangle \cap C = \{1\}$, $\langle b \rangle \cap C = \{1\}$. Show that $\langle a, b \rangle$ is isomorphic to the free product $\langle a \rangle * \langle b \rangle$. [Use the result that no reduced word of length greater than 1 in a free product with amalgamation represents the identity element.]

2.2 Let $G = \langle t, A \mid tBt^{-1} = C \rangle$ be an HNN-extension. Show that t has infinite order in G, and $\langle t \rangle \cap A = \langle 1 \rangle$.

In the group with presentation $\langle a, b, c \mid aba^{-1} = b^2, bcb^{-1} = c^2 \rangle$, show that a and c have infinite order and $\langle a \rangle \cap \langle c \rangle = \langle 1 \rangle$.

3. Graphs and CW-complexes

In this section the reader is assumed to be familiar with the idea of fundamental group of a topological space. See [15, Chapter 2]. A CW-complex is a topological space obtained from a discrete space by successively attaching Euclidean balls of increasing dimension via their boundaries. This will be clarified later. The idea of a graph which will be used is defined immediately below; associated to a graph is a group, defined combinatorially, called its fundamental group, which can be shown to be a free group. Given a graph, there is a one-dimensional CW-complex whose fundamental group is isomorphic to the combinatorial fundamental group. Finally, given a presentation, there is a two-dimensional CW-complex whose fundamental group has the given presentation.

Graphs

The definition of graph follows that in [17].

Definition. A *directed graph* X consists of two disjoint sets, $V(X)$ and $E(X)$ (the set of *vertices* and *edges* of X, respectively), together with two mappings $o, t : E(X) \to V(X)$. The vertices $o(e)$, $t(e)$ are called the *endpoints* of e.

Definition. A *graph* is a directed graph X together with a mapping $E(X) \to E(X)$, $e \mapsto \bar{e}$, satisfying: $e \neq \bar{e}$, $\bar{\bar{e}} = e$, $o(\bar{e}) = t(e)$ and $t(\bar{e}) = o(e)$, for all $e \in E(X)$. An *unoriented edge* of X is a pair $\{e, \bar{e}\}$, where $e \in E(X)$.

The *endpoints* of $\{e, \bar{e}\}$ are those of e, which are the same as those of $\bar{e}$. Following the usual practice, a graph is pictured by drawing a small circle for a vertex and a line for an unoriented edge joining its endpoints. Note that $o(e) = t(e)$ is allowed, when the line becomes a circle.

Definition. An *orientation* of a graph X is a set of edges containing exactly one edge from each unoriented edge $\{e, \bar{e}\}$.

Definition. Let X and Y be graphs. A *graph map* from X to Y is a mapping $f : V(X) \cup E(X) \to V(Y) \cup E(Y)$ which maps vertices to vertices and edges to edges, such that, for all edges $e \in V(X)$, $f(o(e)) = o(f(e))$, $f(t(e)) = t(f(e))$ and $f(\bar{e}) = \overline{f(e)}$.

A graph map is called an *isomorphism* if it is a bijective map.

A graph X is a *subgraph* of a graph Y if $V(X) \subseteq V(Y)$, $E(X) \subseteq E(Y)$ and the inclusion map $V(X) \cup E(X) \hookrightarrow V(Y) \cup E(Y)$ is a graph map, which means that, if $e \in E(X)$ then $o(e)$, $t(e)$ and $\bar{e}$ have the same meaning in Y as they do in X.

Paths. Let L_n ($n \geq 0$) be the graph with vertex set $\{0, 1, \ldots, n\}$ and edge set $\{(0,1), (1,2), \ldots, (n-1, n)\} \cup \{(1,0), (2,1), \ldots, (n, n-1)\}$, with $o(i, i+1) = i$, $t(i, i+1) = i+1$, $\overline{(i, i+1)} = (i+1, i)$.

Definition. A *path* of length n in a graph X is a graph map $p : L_n \to X$. Denote $p(0)$ by $o(p)$ and $p(n)$ by $t(p)$ and call these the *endpoints* of p.

One says that p *joins* $o(p)$ to $t(p)$, and that it is a path from $o(p)$ to $t(p)$.

A path p of length $n > 0$ defines a sequence of edges $(e_1, \ldots, e_n)$, where $e_i = p(i-1, i)$, satisfying $t(e_i) = o(e_{i+1})$ for $1 \leq i \leq n-1$, and any sequence of edges satisfying these conditions defines a path. Note that $o(p) = o(e_1)$ and $t(p) = t(e_n)$. Paths of positive length will be viewed in this way as edge sequences.

Also, for every vertex v of X, there is a *trivial path*, denoted by 1_v, with $o(1_v) = t(1_v) = v$, of length 0.

Definition. A graph X is *connected* if, given $x, y \in V(X)$, there exists a path in X from x to y.

A path $(e_1, \ldots, e_n)$ is called *reduced* if $e_i \neq \bar{e}_{i+1}$ for $1 \leq i \leq n-1$ (trivial paths are regarded as reduced), and a path p is *closed* if $o(p) = t(p)$.
A *circuit* is a closed reduced path of positive length, $(e_1, \ldots, e_n)$ such that $o(e_i) \neq o(e_j)$ for $1 \leq i, j \leq n$ with $i \neq j$.

Definition. A *tree* is a connected graph having no circuits.

Lemma 3.1. *A graph X is a tree if and only if, for all vertices u, v of X, there is a unique reduced path which joins u to v.*

Proof. Part of this is given as an exercise (Exercise 3.1); for a full proof, see [5, Section 5.1]. □

A *subtree* of a graph is a subgraph which is a tree. A *maximal tree* in a graph X is a subtree of X which is not properly contained in any other subtree of X.

Lemma 3.2. *Let X be a connected graph. Then*

(1) *a subtree T of X is maximal if and only if $V(T) = V(X)$;*
(2) *X has a maximal subtree.*

Proof. Again see [5, Section 5.1]. □

The Fundamental group. Given two paths p and q with $t(p) = o(q)$, a product pq can be defined as follows. If $p = (e_1, \ldots, e_n)$ and $q = (f_1, \ldots, f_m)$, then $pq = (e_1, \ldots, e_n, f_1, \ldots, f_m)$. If q is trivial, then $pq = p$, and if p is trivial then $pq = q$. Thus pq is a path from $o(p)$ to $t(q)$.

If $p = (e_1, \ldots, e_n)$ is a path, $\bar{p}$ means the path $(\bar{e}_n, \ldots, \bar{e}_1)$, and $\bar{1}_v$ is defined to be 1_v. Note that, for any path p, $\bar{\bar{p}} = p$, $o(\bar{p}) = t(p)$, $t(\bar{p}) = o(p)$.

Let p, q be paths; $p \simeq q$ means one is obtained from the other by deleting a consecutive pair of edges of the form $e, \bar{e}$.

Definition. Paths p and q are *freely equivalent*, written $p \sim q$, if there is a sequence of paths $p = p_1, p_2, \ldots p_n = q$, where $p_i \simeq p_{i+1}$ for $1 \leq i \leq n-1$.

This is an equivalence relation on the set of all paths; the equivalence class of p is denoted by $[p]$. Note that freely equivalent paths have the same endpoints.

Lemma 3.3. *Any path is equivalent to a unique reduced path.*

Proof. This is just like the proof of Lemma 1.2. □

Lemma 3.4. (1) *If $p \sim p'$, $q \sim q'$ and $t(p) = o(q)$ then $pq \sim p'q'$.*
(2) *If $p \sim q$ then $\bar{p} \sim \bar{q}$.*

Proof. This is left as an exercise. □

Let v_0 be a vertex of the connected graph X. Let $\pi_1(X, v_0)$ be the set of equivalence classes of paths starting and ending at v_0. If $[p], [q] \in \pi_1(X, v_0)$, then by Lemma 3.4(1), one can define their product by $[p][q] := [pq]$. This makes $\pi_1(X, v_0)$ into a group with identity element $[1_{v_0}]$, and $[p]^{-1}$ is $[\bar{p}]$.

Definition. The group $\pi_1(X, v_0)$ is called the *fundamental group* of X at v_0.

Now choose a maximal tree T of X. For $v \in V(X)$, let p_v be the reduced path in T from v_0 to v. For $e \in E(X)$, let $w_e = p_{o(e)} e \overline{p_{t(e)}}$. Then $[w_e] \in \pi_1(X, v_0)$, and $[w_e] = 1$ if $e \in E(T)$. If $e \notin E(T)$ then w_e is reduced. Further,

$$w_{\bar{e}} = p_{o(\bar{e})} \bar{e}\, \overline{p_{t(\bar{e})}} = p_{t(e)} \bar{e}\, \overline{p_{o(e)}} = \overline{w_e},$$

hence $[w_e]^{-1} = [w_{\bar{e}}]$.

Suppose $p = (e_1, \ldots, e_n)$ is a path from v_0 to v_0 in X. Then $w_{e_1} \ldots w_{e_n} \sim p$, so $[p] = [w_{e_1}] \ldots [w_{e_n}]$. Let A be an orientation of X. It follows that $\pi_1(X, v_0)$ is generated by $U = \{[w_e] \mid e \in A \setminus E(T)\}$.

Theorem 3.5. *In fact, $\pi_1(X, v_0)$ is free with basis U.*

Proof. This can be proved by elaborating the proof of [19, Theorem 1.4.2]. $\square$

Cayley Graphs. Let G be a group and let $\alpha : X \to G$ be a mapping such that $\alpha(X)$ generates G. Form a directed graph as follows: the set of vertices is G, and the set of edges is $G \times X$; endpoints are defined by $o(g, x) = g$, $t(g, x) = g\alpha(x)$.

For every edge $e = (g, x)$ add an edge $\bar{e}$, with $o(\bar{e}) = g\alpha(x)$, $t(\bar{e}) = g$. This defines a graph, called the *Cayley graph* of G with respect to α, denoted by $\Gamma(G, \alpha)$. Usually, α is suppressed and the graph is denoted by $\Gamma(G, X)$.

CW-complexes. In this section the reader is assumed to be familiar with the idea of *fundamental group* at a point in a topological space X. As usual, a *path* in X means a continuous map $\alpha : [0, 1] \to X$. If $x = (x_1, \ldots, x_n) \in \mathbb{R}^n$, $\|x\|$ means $(x_1^2 + \cdots + x_n^2)^{1/2}$, the usual Euclidean norm. Let $E^n = \{x \in \mathbb{R}^n \mid \|x\| \leq 1\}$, $U^n = \{x \in \mathbb{R}^n \mid \|x\| < 1\}$, $S^{n-1} = E^n - U^n$ (the closed unit ball, open unit ball and unit $(n-1)$-sphere in $\mathbb{R}^n$, respectively).

Let A be a Hausdorff space, Λ a set; for $\lambda \in \Lambda$, let E_λ be a copy of E^n, so containing copies S_λ, U_λ of S^{n-1}, U^n respectively. Let $f_\lambda : S_\lambda \to A$ be a continuous map, for $\lambda \in \Lambda$. Let $Z = A \amalg \coprod_{\lambda \in \Lambda} E_\lambda$, and let X be the quotient space $Z/\sim$, where $\sim$ is the equivalence relation generated by $z \sim f_\lambda(z)$, $\forall z \in S_\lambda$, $\forall \lambda \in \Lambda$. Let $q : Z \to X$ be the quotient map: q maps A homeomorphically onto a closed subspace of X, which is identified with A via q. Also, X is Hausdorff.

Definition. Let Y be a Hausdorff space having A as a subspace. Then Y is *obtained from A by adjoining n-cells* if there exists a homeomorphism $h : X \to Y$, for some X constructed as above, such that the restriction of h to A is the identity map on A.

Put $p = h \circ q$, let $c_\lambda = p(E_\lambda)$ (the *n-cells*), let $p_\lambda = p|_{E_\lambda}$, and let $f_\lambda = p_\lambda|_{S_\lambda}$ (the p_λ are called the *characteristic maps*, and the f_λ are called the *attaching maps*).

The following facts are noted without proof:

(i) A is closed in Y.
(ii) $p_\lambda|_{U_\lambda}$ is a homeomorphism onto an open subset $\mathring{c}_\lambda$ of Y whose closure in Y is c_λ.
(iii) The $\mathring{c}_\lambda$ are the path components of $Y - A$.
(iv) A subset B of Y is closed if and only if $B \cap A$ is closed in A and $B \cap c_\lambda$ is closed in c_λ, for all $\lambda \in \Lambda$.

Definition. A topological space X is a *CW-complex* if there exist subspaces

$$X^0 \subseteq X^1 \subseteq X^2 \subseteq \cdots \text{ with } X = \bigcup_{n \geq 0} X^n$$

such that each X^n is closed, X^0 is discrete, X^n is obtained from X^{n-1} by adjoining n-cells (for $n \geq 1$), and a subset Y of X is closed if and only if $Y \cap X^n$ is closed in X^n, for all $n \geq 0$.

Note that $E^1 = [-1, 1]$ and $U^1 = (-1, 1)$. Let X be a one-dimensional CW-complex (that is, $X = X^1$) and let c be a 1-cell. If g_1, g_2 are homeomorphisms $U^1 \to \mathring{c}$, define $g_1 \sim g_2$ to mean $g_1^{-1} g_2$ is strictly increasing. (This is an equivalence relation on the set of such homeomorphisms with two equivalence classes, $[g]$ and $[g']$, where $g'(t) = g(-t)$.)

An *oriented* 1-cell is a pair $(c, [g])$ where c is a 1-cell and $[g]$ is one of the equivalence classes. Define $\overline{(c, [g])} = (c, [g'])$. If $\varphi : [-1, 1] \to c$ is the characteristic map, put $g = \varphi|_{(-1,1)}$. Then define $o(c, [g]) = \varphi(-1)$, $t(c, [g]) = \varphi(1)$. This defines a graph Γ, with $V(\Gamma) = X^0$, $E(\Gamma)$ the set of oriented 1-cells.

If c is a 1-cell with characteristic map φ and $g = \varphi|_{(0,1)}$, $e = (c, [g])$, then there is a path α_e in X from $o(e)$ to $t(e)$, $\alpha_e = \varphi \circ h$, where $h : [0, 1] \to [-1, 1]$ is $h(t) = 2t - 1$. Define $\alpha_{\bar{e}}$ to be $\varphi' \circ h$ (where

$\varphi'(t) = \varphi(-t)$), so $\alpha_{\bar{e}}(t) = \alpha_e(1-t)$, hence $\alpha_e \alpha_{\bar{e}}$ is homotopic to the constant path at $o(e)$, using an argument similar to the proof that inverses exist in the fundamental group of a topological space.

If $p = (e_1, \ldots, e_m)$ is a path in Γ, define

$$\theta(p) = \text{the homotopy class of } \alpha_{e_1} \ldots \alpha_{e_m}.$$

(This does not depend on the way the right-hand side is bracketed, by the argument to show that multiplication in the fundamental group is associative, and induction on m.) Also, let $\theta(1_v) =$ the homotopy class of the constant path at v, for $v \in V(\Gamma)$.

It follows that if $p \sim q$ then $\theta(p) = \theta(q)$. Hence, if $v \in V(\Gamma)$, there is an induced map $\psi : \pi_1(\Gamma, v) \to \pi_1(X, v)$, $[p] \mapsto \theta(p)$.

Theorem 3.6. *The map $\psi : \pi_1(\Gamma, v) \to \pi_1(X, v)$ is a group isomorphism, so $\pi_1(X, v)$ is a free group.*

Proof. See [15, Theorem 5.1, Chapter 6]. □

2-complexes. Let X be obtained from A by attaching 2-cells and let $\{c_\lambda \mid \lambda \in \Lambda\}$ be the set of 2-cells, with characteristic maps p_λ. Assume that A is path connected.

Let $g : [0, 1] \to S^1$ be the map $t \mapsto e^{2\pi i t}$. Then $\alpha_\lambda := p_\lambda \circ g$ is a closed path in A. Choose $v \in A$ and a path γ_λ in A from v to $\alpha_\lambda(0)$. Then $\gamma_\lambda \alpha_\lambda \gamma_\lambda^{-1}$ is a closed path at v, so its homotopy class w_λ is in $\pi_1(A, v)$.

Theorem 3.7. *The homomorphism $\pi_1(A, v) \to \pi_1(X, v)$ induced by the inclusion map is onto, and the kernel is the normal subgroup of $\pi_1(A, v)$ generated by $\{w_\lambda \mid \lambda \in \Lambda\}$.*

Proof. See [15, Theorem 2.1, Chapter 7]. (Both this result and Theorem 3.6 depend on the Seifert–van Kampen Theorem, an important tool for calculating fundamental groups.) □

Let $\langle X \mid R \rangle$ be a group presentation, let $K^0 = \{v\}$ and let K^1 be obtained from K^0 by adjoining a set of 1-cells $\{c_x \mid x \in X\}$. There is only one choice for the attaching maps, resulting in a set of circles joined at a single point, often called a "bouquet of circles".

Let p_x be the characteristic map, g_x its restriction to $(-1, 1)$, $e_x = (c_x, g_x)$ the corresponding oriented edge, and define $e_{x^{-1}} = \bar{e}_x$. For an oriented edge e_y, $y \in X^{\pm 1}$, let $\alpha_y = \alpha_{e_y}$ be the corresponding path in K^1

as defined above (preceding Theorem 3.6). (If $y = x^{\pm 1}$, $x \in X$, α_y goes once round c_x in a preferred direction.)

Let F be the free group on X. By Theorems 3.5 and 3.6, $x \mapsto$ the homotopy class of α_x extends to an isomorphism $\Phi : F \to \pi_1(K^1, x)$.

If $r \in R$, say $r = y_1 \ldots y_k$, $y_i \in X^{\pm 1}$, define $\alpha_r = \alpha_{y_1} \ldots \alpha_{y_k}$. Then α_r induces $\beta_r : S^1 \to K^1$ with $\alpha_r = \beta_r \circ g$ (recall: $g(t) = e^{2\pi i t}$ for $t \in [0, 1]$).

Adjoin 2-cells $\{c_r \mid r \in R\}$ to K^1 with β_r as attaching map for c_r, to obtain a CW-complex $K = K(X \mid R)$. By Theorem 3.7, $\pi_1(K, v) \cong \pi_1(K^1, v)/N$, N being the normal subgroup generated by $\{w_r \mid r \in R\}$, where w_r is the homotopy class of α_r, i.e., $w_r = \Phi(r)$.

Hence $\pi_1(K, v) \cong F/N'$, where N' is the normal subgroup of F generated by R, so $\pi_1(K, v)$ has presentation $\langle X \mid R \rangle$ (cf. remark (3) after the definition of presentation). Note that β_r depends on the way that $\alpha_{y_1} \ldots \alpha_{y_k}$ is bracketed, but this conclusion is true regardless of the bracketing.

Examples. (1) $\langle x, y \mid xyx^{-1}y^{-1} \rangle$; here, K is homeomorphic to the torus $S^1 \times S^1$.

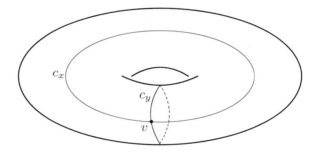

The single attaching map β_r has the effect of identifying certain points on S^1, and the result is the same as identifying points on the sides of a square in the usual way to obtain a torus (cf. [5, Fig. 8, p. 93]); $\pi_1(K, v)$ is the direct product of two infinite cyclic groups.

(2) $\langle x, y \mid xyx^{-1}y \rangle$; similarly, K is the *Klein bottle*. The presentation transforms to $\langle a, b \mid a^2 = b^2 \rangle$ by Tietze transformations.

(3) $\langle x \mid x^2 \rangle$; here K is the *real projective plane* $\mathbb{RP}^2$, obtained by identifying antipodal points of S^2 (cf. [5, p. 52]).

It is possible to realise both free products with amalgamation and HNN-extensions by topological constructions, and from this viewpoint, they are closely related ideas. See [5], Section 4.2, after Corollary 3.

Exercises

3.1 If $p = (e_1, \ldots, e_n)$ is a closed, reduced path of positive length in a graph, show that there exist i, j with $i < j$ such that $(e_i, \ldots, e_j)$ is a circuit. Prove that, if x, y are vertices of a tree, then there is a unique reduced path in the tree from x to y.

3.2 Let X be a set of generators for a group G. Show that the Cayley graph $\Gamma(G, X)$ is connected.

4. The Bass–Serre Theory

This theory describes the structure of a group acting on a tree, as the "fundamental group" of a "graph of groups", a generalisation of free products with amalgamation and HNN-extensions. It is useful in dealing with these groups, and two applications are given. Mostly, proofs will be omitted and the reader is referred to [5, Chapter 8] or [17, Chapter I].

Graphs of groups

Definition. A graph of groups $(\mathcal{G}, Y)$ consists of the following:

- a connected graph Y;
- for all $v \in V(Y)$, a group G_v;
- for all $e \in E(Y)$, a group G_e, with $G_e = G_{\bar{e}}$;
- for all $e \in E(Y)$, a monomorphism $G_e \to G_{t(e)}$, $a \mapsto a^e$.

Let $(\mathcal{G}, Y)$ be a graph of groups, and let T be a maximal tree of Y.

Definition. The *fundamental group* $\pi(\mathcal{G}, Y, T)$ is the group with presentation

$$\langle q_e \ (e \in E(Y)), \ G_v \ (v \in V(Y)) \mid \mathrm{rel}(G_v), q_e a^e q_e^{-1} = a^{\bar{e}} \ (a \in G_e, e \in E(Y)),$$
$$q_e q_{\bar{e}} = 1 \ (e \in E(Y)), q_e = 1 \ (e \in E(T))\rangle.$$

(The set of generators is the disjoint union of $\{q_e \mid e \in E(Y)\}$ and the G_v, for $v \in V(Y)$.)

Examples. (1) $E(Y) = \{e, \bar{e}\}$, $V(Y) = \{o(e), t(e)\}$, $o(e) \neq t(e)$. Let $A = G_{o(e)}$, $B = G_{t(e)}$, $C = G_e = G_{\bar{e}}$. The only maximal tree is $T = Y$, so the presentation is

$$\langle q_e, q_{\bar{e}}, A, B \mid \mathrm{rel}(A), \mathrm{rel}(B),\ q_e c^e q_e^{-1} = c^{\bar{e}},\ q_{\bar{e}} c^{\bar{e}} q_{\bar{e}}^{-1} = c^e\ (c \in C),$$

$$q_e q_{\bar{e}} = 1, q_{\bar{e}} q_e = 1, q_e = 1, q_{\bar{e}} = 1 \rangle.$$

After Tietze transformations, this becomes $\langle A, B \mid \mathrm{rel}(A), \mathrm{rel}(B), c^e = c^{\bar{e}}\ (c \in C) \rangle$, a presentation of $A *_C B$.

(2) $E(Y) = \{e, \bar{e}\}$, but $o(e) = t(e)$ and $V(Y) = \{o(e)\}$. Let $A = G_{o(e)}$. There is only one maximal tree T, with $V(T) = V(Y)$ and $E(T) = \emptyset$. After Tietze transformations, the presentation is

$$\langle t, A \mid \mathrm{rel}(A), ta^e t^{-1} = a^{\bar{e}}\ (a \in G_e) \rangle$$

where $t = q_e$ (an HNN-extension, base A, stable letter t, associative subgroups B, C, where $B = \mathrm{Im}(G_e \to G_{t(e)})$, $C = \mathrm{Im}(G_{\bar{e}} \to G_{t(\bar{e})})$).

(3) Similarly, any HNN-extension can be expressed as the fundamental group of a graph of groups with one vertex, and an unoriented edge for each stable letter.

(4) Let $\alpha_i : A \to G_i$ be a monomorphism, for $i \in I$.
Take $v \notin I$, let $V(Y) = \{v\} \cup I$, let $E(Y) = \{(v, i), (i, v) \mid i \in I\}$, with $o(v, i) = v$, $t(v, i) = i$ and let $\overline{(v, i)} = (i, v)$. This gives a tree Y. Associate G_i to the vertex i, let $G_v = A$ and $G_e = A$ for all edges e.

The monomorphism $A = G_{(v,i)} \to G_i$ is α_i and the monomorphism $G_{(i,v)} \to G_v$ is the identity map. The resulting graph of groups $(\mathcal{G}, Y)$ has $\pi(\mathcal{G}, Y, Y) \cong *_A G_i$.

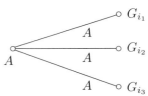

(5) For any connected graph Y, there is a *trivial graph of groups* $(\mathcal{I}, Y)$ with trivial groups assigned to all vertices and edges. If T is a maximal tree of Y, then $\pi(\mathcal{I}, Y, T)$ has presentation

$$\langle q_e\ (e \in E(Y)) \mid q_e q_{\bar{e}} = 1\ (e \in E(Y)),\ q_e = 1\ (e \in E(T)) \rangle$$

a free group with basis $\{q_e \mid e \in A \setminus E(T)\}$, where A is an orientation of Y.

Definition. $F(\mathcal{G}, Y)$ is the group with presentation

$$\langle e, G_v(e \in E(Y), v \in V(Y)) \mid \mathrm{rel}(G_v), ea^e e^{-1} = a^{\bar{e}},$$

$$e\bar{e} = 1\ (a \in G_e, e \in E(Y)) \rangle$$

or $\langle e, G_v(e \in E(Y), v \in V(Y)) \mid \mathrm{rel}(G_v), ea^e \bar{e} = a^{\bar{e}}\ (a \in G_e, e \in E(Y)) \rangle$.

(This is an HNN-extension with base $*_{v \in V(Y)} G_v$, so G_v embeds naturally in $F(\mathcal{G}, Y)$.)

Definition. Let $v_0 \in V(Y)$. Then $\pi(\mathcal{G}, Y, v_0)$ is the subgroup of $F(\mathcal{G}, Y)$ consisting of all elements of $F(\mathcal{G}, Y)$ represented by a word of the form

$$(r_0, e_1, r_1, e_2, r_2, \ldots, e_n, r_n),$$

where $n \geq 0$, $(e_1, \ldots, e_n)$ is a closed path at v_0, $r_0 \in G_{v_0}$ and for $1 \leq i \leq n$, $r_i \in G_{t(e_i)}$.

Let T be a maximal tree in Y. Then there is a homomorphism $\alpha : F(\mathcal{G}, Y) \to \pi(\mathcal{G}, Y, T)$, induced by $e \mapsto q_e$ for $e \in E(Y)$ and $g \mapsto g$ for $g \in G_v$, $v \in V(Y)$.

Lemma 4.1. *Let $v_0 \in V(Y)$. Then α, restricted to $\pi(\mathcal{G}, Y, v_0)$, is an isomorphism onto $\pi(\mathcal{G}, Y, T)$.*

Proof. If p is a path in Y, it is a word in the generators of $F(\mathcal{G}, Y)$, so represents an element g_p of $F(\mathcal{G}, Y)$. For $v \in V(Y)$, let p_v be the reduced path in T from v_0 to v, and let $g_v = g_{p_v}$. The maps

$$G_v \to F(\mathcal{G}, Y), \ g \mapsto g_v g g_v^{-1} \quad \text{and} \quad q_e \mapsto g_{o(e)} e g_{t(e)}^{-1}$$

where $v \in V(Y)$ and $e \in E(Y)$ induce a homomorphism $\beta : \pi(\mathcal{G}, Y, T) \to F(\mathcal{G}, Y)$. Clearly $\text{Im}(\beta) \subseteq \pi(\mathcal{G}, Y, v_0)$, and β and $\alpha|_{\pi(\mathcal{G}, Y, v_0)}$ are inverse mappings. $\square$

Corollary 4.2.

(1) *The group $\pi(\mathcal{G}, Y, v_0)$ is independent, up to isomorphism, of the choice of v_0.*
(2) *The group $\pi(\mathcal{G}, Y, T)$ is independent, up to isomorphism, of the choice of T.*
(3) *The natural maps $G_v \to \pi(\mathcal{G}, Y, T)$, for $v \in V(Y)$, are monomorphisms.*

Proof. Part (1) follows from Lemma 4.1 by fixing T and letting v_0 vary, and (2) follows by fixing v_0 and letting T vary. The map $G_v \to \pi(\mathcal{G}, Y, T)$ is the composite of the natural map $G_v \to F(\mathcal{G}, Y)$ and $\alpha|_{\pi(\mathcal{G}, Y, v)}$, and both are monomorphisms. $\square$

Group actions. Let X be a graph. An isomorphism from X to X is called an *automorphism* of X. The set $\text{Aut}(X)$ of automorphisms of X is a group, and an action of G on X is given by a homomorphism $G \to \text{Aut}(X)$. It consists of actions on both $V(X)$ and $E(X)$, such that $go(e) = o(ge)$, $gt(e) = t(ge)$ and $g\bar{e} = \overline{ge}$ for all $g \in G$ and $e \in E(X)$.

Definition. An action of a group G on a graph X is *without inversions* if $ge \neq \bar{e}$ for all $g \in G$ and $e \in E(X)$.

When an action is without inversions, one can define the *quotient graph* $Y = X/G$ as follows: $V(Y) = $ set of G-orbits of $V(X)$ and $E(Y) = $ set of G-orbits of $E(X)$. Put $o(Ge) = Go(e)$, $t(Ge) = Gt(e)$ and $\overline{Ge} = G\bar{e}$ ($Ge \neq G\bar{e}$ since the action is without inversions).

The map $p: X \to Y$ given by $p(z) = Gz$ ($z \in V(X) \cup E(X)$) is a graph map, the *quotient map*.

Associated graphs of groups. Suppose G is a group acting on a graph X. Choose a maximal tree T of $Y = X/G$. It can be shown, for example using Zorn's lemma, that there is a graph map $j: T \to X$ such that $pj = \text{id}_T$. Thus $j(T)$ is an isomorphic copy of T.

Take an orientation A of Y. Let $e \in A \setminus E(T)$, so $e = G\tilde{e}$ for some $\tilde{e} \in E(X)$.

Then $p(o(\tilde{e})) = o(e) = pj(o(e))$, so $o(\tilde{e})$ and $j(o(e))$ are in the same orbit; take $g \in G$ such that $go(\tilde{e}) = j(o(e))$. Then $o(g\tilde{e}) = jo(e)$.

Define $j(e) = g\tilde{e}$ and $j(\bar{e}) = \overline{j(e)}$, so j is extended to all of Y.

In general, j is no longer a graph map, but $pj = \text{id}_Y$ and $\overline{j(e)} = j(\bar{e})$ for all $e \in E(Y)$.

If $e \in A \setminus E(T)$, $p(j(t(e))) = t(e)$, and $p(t(je)) = t(pj(e)) = t(e)$, so $j(t(e))$ and $t(j(e))$ are in the same orbit; choose $\gamma_e \in G$ such that $\gamma_e jt(e) = t(je)$. Also, put $\gamma_{\bar{e}} = \gamma_e^{-1}$, and $\gamma_e = 1$ for $e \in E(T)$.

Now a graph of groups $(\mathcal{G}, Y)$ can be constructed as follows: define $G_z = \text{stab}(j(z))$ for $z \in V(Y) \cup E(Y)$. The monomorphism $G_e \to G_{t(e)}$ is the inclusion map if $e \in E(T)$ or $e \notin A$, otherwise it is $a \mapsto \gamma_e^{-1} a \gamma_e$. Then

$$\gamma_e a^e \gamma_e^{-1} = a^{\bar{e}} \quad \text{for all } e \in E(Y) \text{ and } a \in G_e, \tag{4.1}$$

$$\gamma_e \gamma_{\bar{e}} = 1 \quad \text{for all } e \in E(Y), \tag{4.2}$$

$$\gamma_e = 1 \quad \text{for all } e \in E(T). \tag{4.3}$$

The endpoints of edges in X are given by:

for $e \in A$: $o(gj(e)) = gj(o(e))$, $\quad t(gj(e)) = g\gamma_e j(t(e))$, (4.4)

for $e \notin A$: $o(gj(e)) = g\gamma_e^{-1} j(o(e))$, $\quad t(gj(e)) = gj(t(e))$, (4.5)

where $g \in G$.

Because of (4.1)–(4.3), there is a homomorphism $\phi: \pi(\mathcal{G}, Y, T) \to G$ induced by the identity mapping on each G_v, for $v \in V(Y)$, and $q_e \mapsto \gamma_e$.

Lemma 4.3. *The homomorphism ϕ just defined is surjective.* □

The "universal covering". Suppose $(\mathcal{G}, Y)$ is a graph of groups and T is a maximal tree of Y. Let $G = \pi(\mathcal{G}, Y, T)$ and choose an orientation A of Y. Then a graph $\tilde{Y} = \tilde{Y}(\mathcal{G}, Y, T, A)$ can be defined as follows. Let

$$V(\tilde{Y}) = \coprod_{v \in V(Y)} G/G_v \quad \text{and} \quad E(\tilde{Y}) = \coprod_{e \in E(Y)} G/G'_e$$

where

$$G'_e = \begin{cases} G_{\bar{e}}^{\bar{e}} & \text{if } e \in A, \\ G_e^e & \text{if } e \notin A, \end{cases}$$

and $G_e^e = \text{Im}(G_e \to G_{t(e)})$. These are disjoint unions, so

$$E(\tilde{Y}) = \bigcup_{e \in E(Y)} (G/G'_e) \times \{e\}, \quad V(\tilde{Y}) = \bigcup_{v \in V(Y)} (G/G_v) \times \{v\}.$$

The group G acts on $V(\tilde{Y})$ and $E(\tilde{Y})$ via left multiplication on the cosets. Put $\tilde{e} = (G'_e, e)$, $\tilde{v} = (G_v, v)$, for $e \in E(Y)$, $v \in V(Y)$. For $e \in A$ and $g \in G$, define

$$o(g\tilde{e}) = \widetilde{go(e)}, \tag{4.6}$$

$$t(g\tilde{e}) = gq_e \widetilde{t(e)}, \tag{4.7}$$

$$\overline{g\tilde{e}} = g\tilde{\bar{e}}. \tag{4.8}$$

This is enough to define $\tilde{Y}$; G acts on $\tilde{Y}$ without inversions and $\tilde{Y}/G$ is isomorphic to Y.

Theorem 4.4. *The graph $\tilde{Y}$ is a tree.* □

The structure theorem. Let G be a group acting without inversions on a connected graph X, let $Y = X/G$ and let $p : X \to Y$ be the quotient map. Choose a maximal tree T and an orientation A of Y, and let $j : T \to X$ be a graph map with $pj = \text{id}_T$.

Extend j to all of Y and choose elements $\gamma_e \in G$ such that $t(j(e)) = \gamma_e j(t(e))$ for $e \in A \setminus E(T)$, to obtain an associated graph of groups $(\mathcal{G}, Y)$.

Let $\pi = \pi(\mathcal{G}, Y, T)$, let $\tilde{Y} = \tilde{Y}(\mathcal{G}, Y, T, A)$ and let $\phi : \pi \to G$ be the surjective group homomorphism in Lemma 4.3.

Then $\text{stab}(j(x)) = \text{stab}(\tilde{x})$ for all $x \in V(Y) \cup E(Y)$, and $\phi|_{\text{stab}(\tilde{x})}$ is the identity map.

It follows that a map $\psi : \tilde{Y} \to X$ can be defined by $\psi(g\tilde{x}) = \phi(g)(j(x))$ for all $g \in \pi$, $x \in V(Y) \cup E(Y)$. Using equations (4.4)–(4.8), ψ is a graph map. Clearly ψ is onto, and $\psi(gx) = \phi(g)\psi(x)$ for all $g \in \pi$, $x \in V(\tilde{Y}) \cup E(\tilde{Y})$.

Theorem 4.5 (Bass–Serre). *In these circumstances, the following are equivalent:*

(a) *X is a tree;*
(b) *ψ is an isomorphism;*
(c) *ϕ is an isomorphism.* □

Applications. An action of a group on a tree (or any graph) X is called *free* if $\text{stab}(z) = 1$ for all $z \in V(X) \cup E(X)$.

Corollary 4.6. *A group G is a free group if and only if it acts freely and without inversions on a tree.*

Proof. Suppose G is free with basis S. Let Y be a graph with one vertex and with $E(Y) = S \cup S^{-1}$, where $\bar{s} = s^{-1}$ for $s \in E(Y)$. Let T be the maximal tree of Y ($V(T) = V(Y)$ and $E(T) = \emptyset$). Take the trivial graph of groups $(\mathcal{J}, Y)$; $\pi(\mathcal{J}, Y, T)$ is a free group with basis $\{q_s \mid s \in S\}$, so is isomorphic to G. Therefore G acts without inversions on the tree $\tilde{Y}$ in Theorem 4.4, and the action is free.

If G acts freely and without inversions on a tree X, an associated graph of groups has the form $(\mathcal{J}, Y)$; by Theorem 4.5, $G \cong \pi(\mathcal{J}, Y, T)$, where T is a maximal tree of Y. If A is an orientation of Y, $\pi(\mathcal{J}, Y, T)$ is a free group, with basis $\{q_e \mid e \in A \setminus E(T)\}$. □

Remark 4.7. The tree $\tilde{Y}$ in first part of the proof of Corollary 4.6 is isomorphic to the Cayley graph $\Gamma(G, S)$, so the Cayley graph is a tree. Conversely, if G is a group generated by a set S, then G acts freely on $\Gamma(G, S)$ by left multiplication on the vertices and the first coordinate of the edges, so if $\Gamma(G, S)$ is a tree, G is free by Corollary 4.6, and closer inspection of the proof shows that G is free with basis S. This can be proved directly without using the Bass–Serre theory, using Exercise 1.1.

Corollary 4.8 (Nielsen–Schreier). *A subgroup of a free group is a free group.*

Proof. Let G be a free group, H a subgroup of G. By Corollary 4.6, G acts freely and without inversions on a tree, H acts freely by restriction, so H is free. □

Corollary 4.9 (Kurosh Subgroup Theorem). *If $G = \bigast_{i \in I} G_i$ is a free product, and H is a subgroup of G, then H has a decomposition*

$$H = F * \bigast_{i \in I} \bigast_{g \in R_i} H \cap gG_ig^{-1},$$

where F is a free group and R_i is some suitable set of representatives for the double cosets $\{HgG_i \mid g \in G\}$.

Proof. Let $(\mathcal{G}, Y)$ be the tree of groups in example (4.4) above, with $A = \{1\}$, so G is isomorphic to $\pi(\mathcal{G}, Y, Y)$. By Theorem 4.4 there is a tree $\tilde{Y}$ on which G acts without inversions, and H acts by restriction. Note that $V(\tilde{Y}) = \left(\coprod_{i \in I} G/G_i \right) \amalg G$.

Let $Z = \tilde{Y}/H$ be the quotient graph, choose a maximal tree T and orientation A of Z and form an associated graph of groups $(\mathcal{H}, Z)$ for the action of H on $\tilde{Y}$, using a map $j : Z \to \tilde{Y}$, as above. By Theorem 4.5, $H \cong \pi(\mathcal{H}, Z, T)$.

All edge groups of $(\mathcal{H}, Z)$ are trivial, so (by Tietze transformations) $\pi(\mathcal{H}, Z, T)$ is a free product $F * \bigast_{v \in V(Z)} H_v$, where H_v is the group associated to v in $(\mathcal{H}, Z)$, and F is free, with basis $\{q_e \mid e \in A \setminus E(T)\}$.

If $j(v) = gG_i$ for some $i \in I$ (more accurately, $j(v) = (gG_i, i)$), then $H_v = H \cap gG_ig^{-1}$, the stabilizer of $j(v)$ in H. For $i \in I$, the vertices $j(v)$ of this form are a set of representatives for the orbits of H acting on G/G_i.

These orbits are in one-to-one correspondence with the double cosets HgG_i ($g \in G$), via the map (H-orbit of gG_i) $\mapsto HgG_i$. The other vertex groups of $(\mathcal{H}, Z)$ are trivial, hence the result. $\square$

5. Some Classes of Groups

This is a brief look at some classes of groups with geometric connections which have been successfully studied. No proofs will be given, and it is left to the reader to find them in the references. Many are associated with *simplicial complexes* which help in their study, and so these are discussed first.

Simplicial complexes

Definition. An *abstract simplicial complex* K consists of a set V and a set S of finite non-empty subsets of V, satisfying

(1) if $v \in V$ then $\{v\} \in S$;
(2) if $\sigma \in S$ and $\tau \subseteq \sigma$, then $\tau \in S$.

The set V is called the set of *vertices*, and S the set of *simplices* of K. If $\sigma \in S$ has $n+1$ elements, n is the *dimension* of σ and σ is called an n-simplex.

Let $W = \prod_{v \in V} \mathbb{R}$, be a real vector space. Identify $v \in V$ with the element of W having 1 in the v-coordinate and 0 in all other coordinates. Give each finite-dimensional linear subspace of W the Euclidean topology. Then give W the *weak* topology: a subset is closed if and only if its intersection with every finite-dimensional subspace is closed.

For a simplex $\sigma = \{v_0, v_1, \ldots, v_n\}$, let $|\sigma|$ be the convex hull of $\{v_0, v_1, \ldots v_n\}$ in W. Let $|K|^0 = V$, and

$$|K|^n = |K|^{n-1} \cup \bigcup \{|\sigma| \mid \sigma \text{ is an } n\text{-simplex of } K\}, \quad (n \geq 1)$$

and let $|K| = \bigcup_{n \geq 0} |K|^n$, so $|K| \subseteq W$. Give $|K|$ the relative topology. Then $|K|^0 \subseteq |K|^1 \subseteq \cdots$ and this gives $|K|$ the structure of a CW-complex. (For more information see [16, Chapter 1, Sections 1–3].)

Inclusion of sets, $\subseteq$, is a partial order on S, and any two elements σ, τ have a greatest lower bound in S (namely $\sigma \cap \tau$). Moreover, for any $\sigma \in S$, the set $\{\tau \in S \mid \tau \subseteq \sigma\}$ is isomorphic to the poset of subsets of $\{1, 2, \ldots, r\}$, for some $r \geq 0$.

Conversely, suppose $(P, \leq)$ is a poset satisfying

(i) any two elements σ, τ have a greatest lower bound in P;
(ii) for any $b \in P$, the set $\{a \in P \mid a \leq b\}$ is isomorphic to the poset of subsets of $\{1, 2, \ldots, r\}$, for some $r \geq 0$.

Then a simplicial complex can be obtained as follows. Let V be the set of rank 1 elements (rank being the integer r in (ii)). For $a \in P$, define $a' = \{v \in V \mid v \leq a\}$, and let $S = \{a' \mid a \in P\}$. The resulting simplicial complex is isomorphic as a poset to P, via $a \mapsto a'$.

Hyperbolic groups. Let (X, d) be a metric space.

Definition. A *geodesic* in (X, d) is the image of an isometry $\alpha : [0, a] \to X$, where $a \geq 0$. It *joins* $\alpha(0)$ to $\alpha(a)$; (X, d) is *geodesic* if there exists a geodesic joining any two points.

A geodesic joining x, y is denoted $[x, y]$, even though it may not be unique.
A *geodesic triangle* $\Delta = \Delta(x, y, z)$ is the union of three geodesics

$$[x, y] \cup [y, z] \cup [z, x].$$

A *comparison triangle* for Δ in $\mathbb{R}^2$ is a triangle $\overline{\Delta}$ in $\mathbb{R}^2$ with vertices $\bar{x}, \bar{y}, \bar{z}$ such that $\|\bar{x} - \bar{y}\| = d(x, y)$, etc. There are isometries $[\bar{x}, \bar{y}] \to [x, y]$ etc., which give a map $p_\Delta : \overline{\Delta} \to \Delta$.

Definition. If $u \in \Delta$, any point $\bar{u}$ such that $p_\Delta(\bar{u}) = u$ is called a *comparison point* for u.

Let c_x, c_y, c_z be the *interior pts* of $\overline{\Delta}$, i.e., the points of contact of the sides with the incircle. Further, let T_Δ be the "tripod" obtained by isometrically identifying $[\bar{x}, c_y]$ and $[\bar{x}, c_z]$, etc.

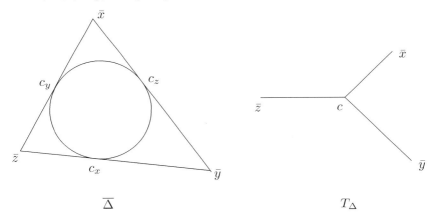

Let $q : \overline{\Delta} \to T_\Delta$ be the quotient map. It is easy to see that there is an induced map $f_\Delta : \Delta \to T_\Delta$, with $f_\Delta \circ p_\Delta = q$.

Definition. Let $\delta \geq 0$; Δ is δ-*thin* if, for all $u, v \in \Delta$, $f_\Delta(u) = f_\Delta(v)$ implies that $d(u, v) \leq \delta$.

Suppose (X, d) is a metric space. Choose a point $z \in X$ and, for $x, y \in X$, define
$$(x \cdot y)_z = \tfrac{1}{2}(d(x, z) + d(y, z) - d(x, y)).$$

(In the comparison triangle above, $(x \cdot y)_z = \|\bar{z} - c_x\| = \|\bar{z} - c_y\|$.)

Definition (Gromov). Let (X, d) be a metric space, let $v \in X$, and let $\delta \in \mathbb{R}$ with $\delta \geq 0$. Then (X, d) is δ-*hyperbolic with respect to* v if
$$\forall x, y, z \in X, \quad x \cdot y \geq \min\{x \cdot z, y \cdot z\} - \delta,$$
where $x \cdot y$ means $(x \cdot y)_v$, etc.

Lemma 5.1. *If (X, d) is δ-hyperbolic with respect to v, and t is any other point of X, then (X, d) is 2δ-hyperbolic with respect to t.*

Definition. A metric space (X, d) is δ-hyperbolic if it is δ-hyperbolic with respect to all points of X, and is hyperbolic if it is δ-hyperbolic for some $\delta \geq 0$.

Proposition 5.2. *Let (X, d) be a geodesic metric space.*

(1) *If (X, d) is δ-hyperbolic, then all geodesic triangles are 4δ-thin.*
(2) *If all geodesic triangles are δ-thin, then (X, d) is δ-hyperbolic.*

Examples.

(1) In hyperbolic space (in the usual sense), triangles are δ-thin; cf. [6, Chapter 1, Section 4].
(2) Complete simply connected Riemannian manifolds with sectional curvature κ satisfying $\kappa \leq c < 0$ for some c are hyperbolic.
(3) A geodesic 0-hyperbolic metric space is an $\mathbb{R}$-*tree*.
(4) A bounded metric space (X, d) is hyperbolic ($\delta = \text{diam}(X)$).

If Γ is a connected graph, the *path metric* d on $V(\Gamma)$ is defined by: $d(x, y)$ is the length of a shortest path joining x and y.

This applies for example, if Γ is a Cayley graph; the path metric on $V(\Gamma(G, X)) = G$ is given by $d(g, h) = L(g^{-1}h)$, where $L(g)$ is the length of a shortest word in $X^{\pm 1}$ representing g.

Definition. Suppose X is a finite set of generators for a group G. Then G is said to be *word hyperbolic* if $(\Gamma(G, X), d)$ is hyperbolic, where d is the path metric on G (the vertex set of $\Gamma(G, X)$).

It can be shown that this is independent of the choice of finite generating set X.

Examples.

(1) Finite groups.
(2) Finitely generated free groups (the Cayley graph is a tree, and trees are 0-hyperbolic).
(3) Surface groups with negative Euler characteristic.
(4) Any group containing a free abelian subgroup of rank 2 is not word hyperbolic.

Definition. Let (X, d) be a metric space; the *Rips complex* $P_k(X)$ is the simplicial complex with vertex set X, where a finite subset is a simplex if and only if its diameter $\leq k$.

Theorem 5.3. *If (X, d) is a geodesic δ-hyperbolic metric space and $k > 0$, then $|P_k(X)|$ is contractible for $k \geq 4\delta$.*

This makes the Rips complex useful in studying the structure of hyperbolic groups.

CAT(0) spaces.

Definition. Let Δ be a geodesic triangle in (X, d); Δ is CAT(0) if, for all $u, v \in \Delta$, and comparison points $\bar{u}, \bar{v}$ for u, v respectively, $d(u, v) \leq \|\bar{u} - \bar{v}\|$. The space (X, d) is CAT(0) if it is geodesic and all triangles are CAT(0).

It can be shown that CAT(0) spaces are contractible. A group is CAT(0) if it has a proper, cocompact action on a CAT(0) space. (A group action on metric space (X, d) is *proper* if for all $x \in X$, there exists $r > 0$ such that $\{g \in G \mid gB(x,r) \cap B(x,r) \neq \emptyset\}$ is finite, where $B(x,r)$ is the open ball $\{y \in X \mid d(x,y) < r\}$. The action is *cocompact* if there exists a compact subset K of X such that $X = GK$.)

Coxeter groups. Let I be a set. For $i, j \in I$ let $m_{ij} \in \{1, 2, 3, \ldots\} \cup \{\infty\}$. Suppose $m_{ij} = m_{ji} \geq 2$, for all $i \neq j$, and $m_{ii} = 1$ for all i.

Definition. The corresponding Coxeter group is the group W with presentation
$$\langle s_i \ (i \in I) \mid (s_i s_j)^{m_{ij}} = 1 \ (i, j \in I) \rangle.$$

(If $m_{ij} = \infty$, the relation is omitted.)

Note that for $i = j$, the relation is $s_i^2 = 1$. Also, $(s_j s_i)^{m_{ji}} = 1$ is a consequence of $(s_i s_j)^{m_{ij}} = 1$. Put $S = \{s_i \mid i \in I\}$; (W, S) is called a *Coxeter system*.

Examples.

(1) The dihedral group D_n, for $n \geq 2$, with presentation $\langle y, u \mid y^2 = 1, u^2 = 1, (yu)^n = 1 \rangle$.
(2) The infinite dihedral group $C_2 * C_2$.
(3) The symmetric group S_n, for $n \geq 2$ (see example (3), p. 90).

Let V be an $\mathbb{R}$-vector space with basis $\{e_i \mid i \in I\}$. Define a symmetric bilinear form $(-, -)$ on V by $(e_i, e_j) = -\cos(\pi/m_{ij})$. Let σ_i be the "reflection" $\sigma_i(v) = v - 2(v, e_i)e_i$. Then $\sigma_i^2 = 1$, and σ_i is clearly linear, so $\sigma_i \in \mathrm{GL}(V)$. Also, $\sigma_i(e_i) = -e_i$, so σ_i has order 2. One can also show that $\sigma_i \sigma_j$ has order m_{ij}. It follows that the map $s_i \mapsto \sigma_i$ induces a homomorphism $W \to \mathrm{GL}(V)$. Hence s_i has order 2 in W and $s_i s_j$ has order m_{ij}.

Theorem 5.4. *This homomorphism is injective.*

If $T \subseteq S$, say $T = \{s_i \mid i \in J\}$, where $J \subseteq I$, let $W_T = \langle T \rangle$.

Theorem 5.5. *The group W_T is itself a Coxeter group, and (W_T, T) is a Coxeter system with the m_{ij} the same as in (W, S), for $i, j \in J$.*

Let (W, S) be a Coxeter system, with I finite. A *special coset* is a coset of W of the form wW_T, where $w \in W$ and $T \subseteq S$. Order the set of special cosets by: $A \leq B$ if and only if $A \supseteq B$. This poset satisfies the conditions (i) and (ii) above, so defines a simplicial complex $\Sigma(W, S)$, called the *Coxeter complex* of (W, S).

There is a way of associating a simplicial complex to any poset $(P, \leq)$; the *flag complex* of $(P, \leq)$, has vertex set P and n-simplices all chains (totally ordered subsets) with $n + 1$ elements.

Let $P = \{wW_T \mid w \in W, T \subseteq S, W_T \text{ finite}\}$, ordered by $\subseteq$. The corresponding simplicial complex is called the *Davis complex* of (W, S), denoted D_W. The group W acts on D_W, hence on $|D_W|$, by left multiplication.

Theorem 5.6 (Davis). $|D_W|$ *is contractible.*

Theorem 5.7 (Moussong). *In fact, $|D_W|$ has a natural $CAT(0)$ metric.*

Artin groups. Given a Coxeter system (W, S), using Tietze transformations, $(s_i s_j)^{m_{ij}} = 1$ $(i \neq j)$ can be rewritten as

$$\underbrace{s_i s_j s_i \ldots}_{m_{ij}} = \underbrace{s_j^{-1} s_i^{-1} s_j^{-1} \ldots}_{m_{ij}}$$

and since $s_i^2 = 1$, as

$$\underbrace{s_i s_j s_i \ldots}_{m_{ij}} = \underbrace{s_j s_i s_j \ldots}_{m_{ij}}.$$

Call this relation r_{ij}. Now omit the relations $s_i^2 = 1$ to obtain

$$\langle S \mid r_{ij} \ (i, j \in I, i \neq j) \rangle.$$

Definition. The group A with this presentation is called the *Artin group* associated to (W, S).

There is a homomorphism $A \to W$ induced by $s_i \mapsto s_i$.

There are two analogues of the Davis complex for Artin groups: the *Deligne complex* and the *Salvetti complex*; A acts freely on the Salvetti complex.

A well-studied special case is the following: A is *right angled* if, for all $i \neq j$, m_{ij} is equal to either 2 or ∞. The relations then say that certain pairs of generators commute. In particular, finitely generated free groups (all $m_{ij} = \infty$) and finitely generated free abelian groups, that is, direct products of finitely many infinite cyclic groups, are Artin groups. (The latter is the case $m_{ij} = 2$ for $i \neq j$, cf. [11, Section 6, Chapter 2].)

Braid groups. Let M be a connected manifold of dimension at least 2. Define

$$F_n(M) = \{(x_1, \ldots, x_n) \in M \times \cdots \times M \mid x_i \neq x_j \text{ for } i \neq j\}.$$

Then $F_n(M)$ is path connected. The symmetric group S_n acts on $F_n(M)$ via: $\sigma(x_1, \ldots, x_n) = (x_{\sigma(1)}, \ldots, x_{\sigma(n)})$ for $\sigma \in S_n$. Let $C_n(M) = F_n(M)/S_n$ (the set of orbits, given the quotient topology), and let $p : F_n(M) \to C_n(M)$ be the quotient map. Choose $c_0 \in C_n(M)$.

Definition. [Fox] The *braid group* $B_n(M)$ is $\pi_1(C_n(M), c_0)$.

(It is independent of c_0 as $C_n(M)$ is path connected.)

Let $\beta \in \pi_1(C_n(M), c_0)$ be represented by $f : [0, 1] \to C_n(M)$, so $f(0) = f(1) = c_0$. Choose $\tilde{c}_0 \in F_n(M)$ such that $p(\tilde{c}_0) = c_0$; p is a *covering map*, so there is a unique path $\tilde{f} : [0, 1] \to F_n(M)$ such that $p \circ \tilde{f} = f$ and $\tilde{f}(0) = \tilde{c}_0$.

Then $\tilde{f}(t) = (\tilde{f}_1(t), \ldots, \tilde{f}_n(t))$, where $\tilde{f}_i : [0, 1] \to M$ is a path in M.

Lemma 5.8.

(i) $\tilde{f}_i(t) \neq \tilde{f}_j(t)$ for $i \neq j$ and $t \in [0, 1]$;
(ii) $\tilde{f}_i(1) = \tilde{f}_{\tau(i)}(0)$ for $1 \leq i \leq n$, for some $\tau \in S_n$.

Define $\alpha_i : [0, 1] \to M \times [0, 1]$ by $\alpha_i(t) = (\tilde{f}_i(t), t)$.

Definition. $\mathcal{A} := (\alpha_1, \ldots, \alpha_n)$ is an *n-string braid* in $M \times [0, 1]$.

Denote the permutation τ by $\tau_{\mathcal{A}}$.

Now specialise to $M = \mathbb{R}^2$; $B_n(\mathbb{R}^2)$ is denoted by B_n, the *braid group on n strings*. Take $\tilde{c}_0 = (x_1, \ldots, x_n)$, where $x_i = (i, 0)$. An n-string braid looks like:

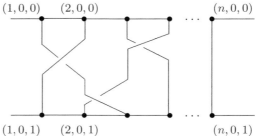

(The strings run between the planes $z = 0$ and $z = 1$, where the z-axis points down.)

In this example, $\tau_{\mathcal{A}}(1) = 3$, $\tau_{\mathcal{A}}(2) = 1$, $\tau_{\mathcal{A}}(3) = 4$ etc.

Corresponding to homotopy of paths in $C_n(\mathbb{R}^2)$ there is a notion of homotopy of braids.

Multiplication in $\pi_1(C_n(\mathbb{R}^2), c_0)$ ($c_0 = p(\tilde{c}_0)$) corresponds to the following operation. If $\mathcal{A}_1$, $\mathcal{A}_2$ are n-string braids, move $\mathcal{A}_2$ under $\mathcal{A}_1$ so the top plane of $\mathcal{A}_2$ is the bottom plane of $\mathcal{A}_1$ and the endpoints match up. Then squeeze this system so it lies between $z = 0$ and $z = 1$. Define $\mathcal{A}_1 \mathcal{A}_2$ to be the resulting braid. (This is Artin's original definition of B_n.) For example

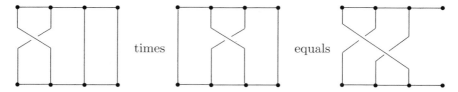

Note that $\tau_{\mathcal{A}_1 \mathcal{A}_2} = \tau_{\mathcal{A}_1} \tau_{\mathcal{A}_2}$. The inverse of a braid is its mirror image with respect to a horizontal plane between $z = 0$ and $z = 1$. Let σ_i be the n-string braid:

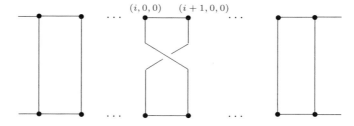

whose inverse is obtained by reversing the twist. Then the homotopy classes of $\sigma_1, \ldots, \sigma_{n-1}$ generate B_n; to see this, make sure every crossing in a braid occurs at a different level, then separate the crossings by horizontal planes.

Theorem 5.9 (Artin). *The braid group B_n has the presentation*

$$\langle \sigma_1, \ldots, \sigma_{n-1} \mid \sigma_i \sigma_j = \sigma_j \sigma_i \ (|i-j| \geq 2), \sigma_i \sigma_{i+1} \sigma_i$$
$$= \sigma_{i+1} \sigma_i \sigma_{i+1} \ (1 \leq i \leq n-2) \rangle.$$

Thus B_n is an Artin group. The corresponding Coxeter group is S_n and the homomorphism $B_n \to S_n$ sends σ_i to the transposition $(i, i+1)$ (see example 3, p. 90). This is the permutation τ_{σ_i}, hence the homomorphism sends any element σ of B_n to τ_σ.

6. Further Reading

A nice introduction to more on the basic theory of free groups and presentations is [11]; there is also an earlier version, volume 22 in the same series (London Mathematical Society Lecture Note Series). For further reading on this topic, and on free products with amalgamation and HNN-extensions, see [12, Chapters 1, 2 and 4], [5, Chapter 1] and [14, Chapters 1–4]. An excellent introduction to the connections with CW-complexes is [15]; this topic is discussed extensively in [12] (but [5] adopts a combinatorial approach).

The Bass–Serre theory is covered in [17, Chapter 1] and [5, Chapter 8]. Of the various classes of groups considered in Section 5, hyperbolic groups are covered in [6, 9, 18]. A very good book for both hyperbolic spaces and CAT(0) spaces is [3]. For an introduction to Coxeter groups see [10] or [7] (the latter also discusses free groups and presentations in detail). For an extensive discussion of Coxeter groups and Artin groups, see [8]. Finally, the standard reference on braid groups is [2], but it does not discuss more recent progress, such as orderability results and linearity of the braid groups.

Solutions to Exercises

1.1 Let F be the free group on X constructed in Section 1 (preceding Exercise 1.2), and let $f : F \to G$ be the extension of the inclusion mapping $X \to G$ to a homomorphism.

Assume G is free with basis X. Then X generates G by the note after the definition of free group. Let $\alpha : X \to F$ be the inclusion map,

$\beta : G \to F$ the extension of α to a homomorphism. Then if $g \in G$ is represented by the non-empty reduced word u, $\beta(g) = u \neq 1$ by the Normal Form Theorem (1.2), so $g \neq 1$.

Conversely, assume X generates G, and no non-empty reduced word in $X^{\pm 1}$ represents the identity element of G. Then f is onto and $f : F \to G$ has trivial kernel, so is an isomorphism. Suppose $\alpha : X \to H$ is a mapping, where G is a group. Then α has a unique extension to a homomorphism $\tilde{\alpha} : F \to H$. Then αf^{-1} is the unique extension of α to a homomorphism $G \to H$, hence G is free with basis X.

1.2 Take a reduced word in $(Y^{\pm 1})$, say $(u_1 \ldots u_n)$, where $u_j = x^{i_j} y^{e_j} x^{-i_j}$ ($i_j \in \mathbb{N}$, $e_j = \pm 1$), $n \geq 1$, which represents an element g of F. Then $g = x^{i_1} y^{e_1} v_1 y^{e_2} v_2 \ldots v_{n-1} y^{e_n} x^{-i_n}$, where $v_j = x^{i_{j+1} - i_j}$ is a power of x. This follows by induction on n. If $v_j = 1$ and $e_j = -e_{j+1}$, then $u_j = u_{j+1}^{-1}$, which is impossible as $(u_1 \ldots u_n)$ is reduced. Hence g is represented by a reduced word in $\{x, y\}^{\pm 1}$ of length at least n, so $g \neq 1$. By Exercise 1.1, $\langle Y \rangle$ is free with basis Y.

1.3 (a) It follows that $yx = x^2 y$ and from $xyx^{-1} = y^2$ we obtain $xy^2 = x^2 yx^{-1} = y$, hence $xy = 1$, so $yx = x^2 y$ implies $yx = x$, hence $y = 1$. Then $xy = 1$ implies $x = 1$.

(b) Firstly, $x^2 y x^{-2} = xy^2 x^{-1} = y^4$, hence

$$y^4 = (zxz^{-1})y(zx^{-1}z^{-1}) = zx(zy)x^{-1}z^{-1}$$
$$= z(xzx^{-1})(xyx^{-1})z^{-1} = z(xzx^{-1})y^2 z^{-1} = zx^{-1}zy^2 z^{-1}$$
$$= zx^{-1}z^{-1}z^2 y^2 z^{-1} = x^{-2} z^2 y^2 z^{-1}$$

so $y^2 = x^{-2} z^2 y^2 z^{-1} y^{-2} = x^{-2} z^2 z^{-4} = x^{-2} z^{-2}$, giving $x^2 y^2 z^2 = 1$. It follows that $x^2(xyx^{-1})z^2 = 1$, so $x^3 y(x^{-1}z^2) = 1$, hence $x^3 yzx^{-1} = 1$, which implies $x^2 yz = 1$. Together with $x^2 y^2 z^2 = 1$, this gives $y^2 z^2 = yz$, so $yz = 1$. Then from $x^2 yz = 1$ we obtain $x^2 = 1$, so $zxz^{-1} = x^2 = 1$, hence $x = 1$. Now $xyx^{-1} = y^2$ implies $y = 1$, and then $yzy^{-1} = z^2$ implies $z = 1$ (or just use the observation that the relations are unchanged by cyclic permutation to conclude $y = z = 1$).

1.4 (a) $\langle a, b \mid aba = bab \rangle \xrightarrow{T1} \langle a, b, y \mid aba = bab, y = ab \rangle$
$\xrightarrow{T3} \langle a, b, y \mid aba = bab, y = ab, y^2 = byb, a = yb^{-1} \rangle$
$\xrightarrow{T4} \langle a, b, y \mid y^2 = byb, a = yb^{-1} \rangle \xrightarrow{T2} \langle b, y \mid y^2 = byb \rangle$
$\xrightarrow{T1} \langle b, y, x \mid y^2 = byb, x = by \rangle$

$\xrightarrow{T3} \langle b, y, x \mid y^2 = byb,\ x = by,\ y^3 = x^2,\ b = xy^{-1}\rangle$
$\xrightarrow{T4} \langle b, y, x \mid y^3 = x^2,\ b = xy^{-1}\rangle \xrightarrow{T2} \langle y, x \mid y^3 = x^2\rangle$
and this can be rewritten as $\langle x, y \mid x^2 = y^3\rangle$.

(b) $\langle a, b, c, d \mid ab = c,\ bc = d,\ cd = a,\ da = b\rangle$
$\xrightarrow{T3\&T4} \langle a, b, c, d \mid ada = c,\ dac = d,\ cd = a,\ da = b\rangle$
$\xrightarrow{T2} \langle a, c, d \mid ada = c,\ dac = d,\ cd = a\rangle$
$\xrightarrow{T3\&T4} \langle a, c, d \mid ada = c,\ c = a^{-1},\ cd = a\rangle$
$\xrightarrow{T3\&T4} \langle a, c, d \mid ada^2 = 1,\ c = a^{-1},\ d = a^2\rangle$
$\xrightarrow{T2} \langle a, d \mid ada^2 = 1,\ d = a^2\rangle \xrightarrow{T3\&T4} \langle a, d \mid a^5 = 1,\ d = a^2\rangle$
$\xrightarrow{T2} \langle a \mid a^5 = 1\rangle$. (This group is one of the *Fibonacci groups*.)

2.1 The inclusion maps $\langle a\rangle \to \langle a, b\rangle$ and $\langle b\rangle \to \langle a, b\rangle$ have an extension to a homomorphism $f : \langle a\rangle * \langle b\rangle \to \langle a, b\rangle$, whose image contains $a = f(a)$ and $b = f(b)$, so f is surjective. Let $g \in \langle a\rangle * \langle b\rangle$ be represented by the reduced word $(g_1, \ldots, g_k)$, where $k \geq 1$. Thus g_i is a non-trivial power of a or of b and g_i, g_{i+1} are not both powers of a or powers of b. Since $\langle a\rangle \cap C = \{1\} = \langle b\rangle \cap C$, this is a reduced word for the free product with amalgamation $A *_C B$, and represents $f(g)$, hence $f(g) \neq 1$. Thus $\ker(f) = \{1\}$ and f is an isomorphism.

2.2 If $n > 0$, then $(1, t, 1, t, \ldots, t, 1)$ (with n entries equal to t) is a reduced word representing t^n, so $t^n \neq 1$ by Britton's lemma. Hence t has infinite order. If $\langle t\rangle \cap A \neq \{1\}$, then there is $n > 0$ such that $a := t^n \in A$. Then $(1, t, 1, t, \ldots, t, a^{-1})$ (with n occurrences of t) is a reduced word representing 1, contradicting Britton's lemma, hence $\langle t\rangle \cap A = \{1\}$

Let H be the group with presentation $\langle b, c \mid bcb^{-1} = c^2\rangle$. Then H is an HNN-extension, base an infinite cyclic group generated by c, stable letter b and associated pair $\langle c\rangle$, $\langle c^2\rangle$. By Corollary 2.4(1), $\langle c\rangle$ may be viewed as a subgroup of H, so c has infinite order in H, and b has infinite order in H by the first part.

The group G with presentation $\langle a, b, c \mid aba^{-1} = b^2, bcb^{-1} = c^2\rangle$ is thus an HNN-extension, base H, stable letter a and associated pair the infinite cyclic subgroups $\langle b\rangle$, $\langle b^2\rangle$ of H. Again H can be viewed as a subgroup of G, so c has infinite order in G, the stable letter a has infinite order by the first part, and since $\langle c\rangle \leq H$, $\langle a\rangle \cap \langle c\rangle = \{1\}$, again by the first part.

3.1 Let $p = (e_1, \ldots, e_n)$ be a closed reduced path of positive length. Then $o(e_1) = t(e_n)$. Choose i, j with $1 \leq i < j \leq n$ such that $o(e_i) = t(e_j)$

and $j - i$ is as small as possible. If $o(e_r) = o(e_s)$ with $i \leq r, s \leq j$ and $r \neq s$, say $r < s$, then $(e_r, \ldots, e_{s-1})$ satisfies $o(e_r) = t(e_{s-1})$, and $(s - 1) - r < s - r \leq j - i$, contradicting minimality of $j - i$. Hence $(e_i, \ldots, e_j)$ is a circuit.

If x, y are vertices of a tree, then a path of minimal length from x to y is reduced. Suppose $p = (e_1, \ldots, e_n)$, $q = (f_1, \ldots, f_m)$ are reduced paths from x to y (which could be trivial). Show by induction on $m+n$ that $p = q$. Now $p\bar{q}$ is a closed path; if it is trivial, then $p = q = 1_x$, so assume it is not trivial. By what has been proved, $p\bar{q}$ is not reduced. This can only happen if $e_n = f_m$, and then $(e_1, \ldots, e_{n-1})$, $(f_1, \ldots, f_{m-1})$ are reduced paths from x to $o(e_n)$. It follows by induction that these paths are equal, hence $p = q$.

3.2 Recall that the edge (g, x), where $g \in G$ and $x \in X$, has label x, and $\overline{(g, x)}$ has label x^{-1}. At each vertex $g \in G$, for every $y \in X^{\pm 1}$, there is an edge with $o(e) = g$ and label y ((g, y) if $y \in X$, and $\overline{(gy, y^{-1})}$ if $y \in X^{-1}$), and $t(e) = gy$. Let $g \in G$, so $g = y_1 \ldots y_n$ for some n and $y_i \in X^{\pm 1}$. Let e_i be an edge starting at $y_1 \ldots y_{i-1}$ with label y_i. Then $(e_1, \ldots, e_n)$ is a path in $\Gamma(G, X)$ from 1 to g. It follows that $\Gamma(G, X)$ is connected.

References

[1] G. Baumslag and D. Solitar, Some two-generator one-relator non-Hopfian groups. *Bull. Amer. Math. Soc.* **68**, 199–201 (1962).
[2] J. S. Birman, *Braids, Links, and Mapping Class Groups*. Annals of Mathematics Studies, Vol. 82. Based on Lecture Notes by James Cannon. Princeton University Press, Princeton, NJ; University of Tokyo Press, Tokyo (1975).
[3] M. R. Bridson and A. Haefliger, *Metric Spaces of Non-positive Curvature*. Grundlehren der Mathematischen Wissenschaften, Vol. 319. Springer-Verlag, Berlin (1999).
[4] I. M. Chiswell, *A Course in Formal Languages, Automata and Groups*. Universitext. Springer-Verlag, London (2009).
[5] D. E. Cohen, *Combinatorial Group Theory: A Topological Approach*, London Mathematical Society Student Texts, Vol. 14. Cambridge University Press (1989).
[6] M. Coornaert, T. Delzant and A. Papadopoulos, *Géométrie et théorie des groupes*, Lecture Notes in Mathematics, Vol. 1441. Springer, Berlin (1990).
[7] H. S. M. Coxeter and W. O. J. Moser, *Generators and Relations for Discrete Groups*, 4th edn., Ergebnisse der Mathematik und ihrer Grenzgebiete, Vol. 14. Springer-Verlag, Berlin (1980).

[8] M. W. Davis, *The Geometry and Topology of Coxeter Groups*. London Mathematical Society Monographs Series, Vol. 32. Princeton University Press, Princeton, NJ (2008).
[9] E. Ghys and P. de la Harpe, *Sur les groupes hyperboliques d'après Mikhael Gromov*. Birkhäuser, Boston (1990).
[10] J. E. Humphreys, *Reflection Groups and Coxeter Groups*. Cambridge Studies in Advanced Mathematics, Vol. 29. Cambridge University Press, Cambridge (1990).
[11] D. L. Johnson, *Topics in the Theory of Group Presentations*. London Mathematical Society Lecture Note Series, Vol. 42. Cambridge University Press, Cambridge (1980).
[12] R. C. Lyndon and P. E. Schupp, *Combinatorial Group Theory*, Ergebnisse der Mathematik und ihrer Grenzgebiete, Vol. 89. Springer, Berlin (1977).
[13] W. Magnus, *Noneuclidean Tesselations and Their Groups*. Academic Press, New York (1974).
[14] W. Magnus, A. Karrass and D. Solitar, *Combinatorial Group Theory. Presentations of Groups in Terms of Generators and Relations*, Reprint of the 1976 2nd edn. Dover Publications, Mineola, NY (2004).
[15] W. S. Massey, *Algebraic Topology: An Introduction*. Harcourt Brace and World, New York (1967).
[16] J. R. Munkres, *Elements of Algebraic Topology*. Addison-Wesley, Menlo Park (1984).
[17] J.-P. Serre, *Trees*. Springer, New York (1980).
[18] H. Short (ed.), Notes on word hyperbolic groups. In *Group Theory from a Geometrical Viewpoint*, eds. E. Ghys, A. Haefliger and A. Verjovsky. World Scientific, Singapore (1991).
[19] H. Zieschang, E. Vogt and H.-D. Coldewey, *Surfaces and Planar Discontinuous Groups*. Lecture Notes in Mathematics, Vol. 835. Springer, Berlin (1980).

Chapter 4

Holomorphic Dynamics and Hyperbolic Geometry

Shaun Bullett

School of Mathematical Sciences
Queen Mary University of London, London E1 4NS, UK
s.r.bullett@qmul.ac.uk

Rational functions are maps of the Riemann sphere to itself of the form $z \to p(z)/q(z)$ where $p(z)$ and $q(z)$ are polynomials. Kleinian groups are discrete subgroups of $\text{PSL}(2,\mathbb{C})$, acting as isometries of three-dimensional hyperbolic space and as conformal automorphisms of its boundary, the Riemann sphere. Work on both experienced remarkable advances in the last two decades of the 20th century and they are the subject of very active continuing research. The aim of this chapter is to introduce the basic ideas in the two areas, emphasising the parallels between them.

1. Introduction

1.1. *Overview*

In this chapter we introduce some of the main themes in the study of iterated *rational functions*, maps of the Riemann sphere $\hat{\mathbb{C}} = \mathbb{C} \cup \{\infty\}$ to itself of the form
$$z \to \frac{p(z)}{q(z)},$$
where p and q are polynomials with complex coefficients, and the study of *Kleinian groups*, discrete groups of maps $\hat{\mathbb{C}} \to \hat{\mathbb{C}}$, each of the form
$$z \to \frac{az+b}{cz+d},$$
where a, b, c, d are complex numbers, with $ad - bc \neq 0$.

There are tantalising similarities between these two areas: results for each frequently suggest what we should look for in the other, a correspondence known as *Sullivan's Dictionary*.

The study of iterated rational maps had its first great flowering with the work of the French mathematicians Julia and Fatou around 1918–1920 (see [2]), though its origins lie earlier, in the late 19th century, in the work of Schottky, Poincaré, Fricke and Klein. It has flowered again spectacularly over the last 35 years, motivated in part by the computer pictures which started to appear from about 1980 onwards, in part by the explosive growth in the subject of chaotic dynamics which started about the same time, and not least by the revolution in three-dimensional hyperbolic geometry initiated by Thurston in the early 1980s. Rational maps and Kleinian groups are very active areas of continuing research: there are major conjectures still unproved. Moreover the remarkable synthesis of complex analysis, hyperbolic geometry and symbolic dynamics that constitutes the subject of holomorphic dynamics has yielded powerful methods for problems which at first sight might appear only to concern the mathematics of real numbers and functions. For example the first conceptual proof of the universality of the Feigenbaum ratios for period doubling renormalisation of real unimodal maps was that of Sullivan in 1992 using holomorphic methods (see [11]).

In a short exposition one cannot hope to include all proofs, and as we proceed we will increasingly refer the reader elsewhere for details. We start by examining examples: formal definitions will come later.

1.2. The family of maps $q_c : z \to z^2 + c$

Figure 1 shows three examples with c real. When $c = 0$ (middle picture) any orbit started inside the unit circle converges to the point 0, any orbit started outside the unit circle heads towards ∞, and any orbit started on the unit circle remains there. The two components of $\{z : |z| \neq 1\}$ are known as the *Fatou set* of the map and the circle $|z| = 1$ is called the *Julia set*. On the unit circle the dynamical behaviour is that of the *(binary) shift*: if we parametrise the circle by $t = \arg(z)/2\pi \in [0, 1) \subset \mathbb{R}$, then $z \to z^2$ becomes $t \to 2t$ mod 1. The periodic points on the circle are those corresponding to t of the form $m/(2^n - 1)$ (exercise). These form a dense set on the unit circle. Moreover the map has *sensitive dependence on initial conditions* on the unit circle, since it doubles distance there.

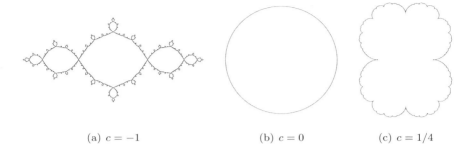

Fig. 1. Julia sets of q_c for three real values of c.

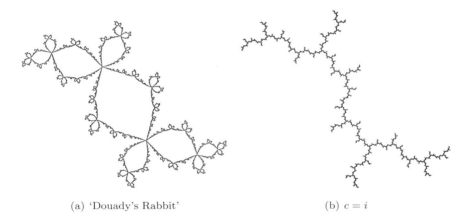

Fig. 2. Julia sets of q_c for two non-real values of c.

In the plot for $z \to z^2 + 1/4$ the Julia set is still a topological circle, and the dynamical behaviour there is still that of the shift, but orbits in the bounded component of the Fatou set now have (forward and backward) limit the *parabolic* fixed point $z = 1/2$, which lies on the Julia set.

When $c = -1$ the Fatou set has infinitely many components. Every orbit started in the outer component is attracted to the fixed point at ∞, but an orbit started in any other component is attracted to the periodic 2-cycle $0 \leftrightarrow -1$. Combinatorially, the Julia set is a *quotient* of the circle, and the dynamics on it is that of the corresponding *quotient* of the shift.

Two plots for non-real c are shown in Fig. 2. The first is for $c = -0.123 + 0.745i$, for which the critical point $z = 0$ has period 3, and the second for $c = i$, for which $z = 0$ is *preperiodic*. It can be proved that whenever the critical point $z = 0$ of q_c is preperiodic but not periodic, the Julia set is a *dendrite* (a connected, simply-connected set with empty interior). A simpler example where 0 is preperiodic is $c = -2$: then the Julia set is the real interval $[-2, 2]$ (Exercise 2, Section 9). For $|c|$ large the Julia set becomes disconnected, indeed a Cantor set. The set of $c \in \mathbb{C}$ for which the Julia set of q_c is connected is known as the *Mandelbrot Set* (see Section 6).

1.3. *Examples of Kleinian groups*

The *modular group* $\mathrm{PSL}(2, \mathbb{Z})$ (see the chapter by Harvey in this volume) maps the open upper half $\mathcal{H}_+$ of the complex plane to itself, the open lower

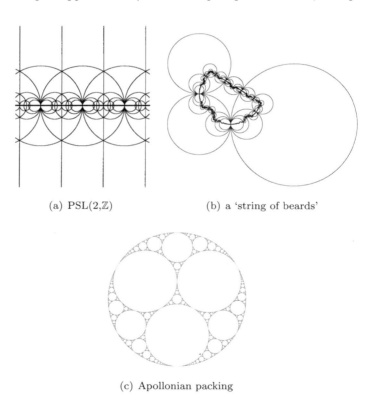

(a) PSL(2,$\mathbb{Z}$)   (b) a 'string of beards'

(c) Apollonian packing

Fig. 3. Limit sets of Kleinian groups.

half-plane $\mathcal{H}_-$ to itself, and the extended real axis $\hat{\mathbb{R}} = \mathbb{R} \cup \infty$ to itself. In Fig. 3(a) we see how $\mathcal{H}_+$ is 'tiled' by the translates of a *fundamental domain* Δ under the elements of the group, and $\mathcal{H}_-$ is tiled by the translates of the mirror image of Δ. Both sets of tiles accumulate on $\hat{\mathbb{R}}$. Just as is the case for rational maps, the action of a Kleinian group G partitions the Riemann sphere into two disjoint completely invariant subsets, an *ordinary set* $\Omega(G)$, and a *limit set* $\Lambda(G)$ (in this case $\hat{\mathbb{R}}$) on which the system exhibits *sensitive dependence on initial conditions*.

A second example is the 'string of beads' illustrated in Fig. 3(b): we start with four circles touching in pairs, and iterate by reflecting (inverting) in each. The reflections reverse orientation, but products of even numbers of these are elements of $\mathrm{PSL}(2,\mathbb{C})$ and form a Kleinian group. The limit set is a topological circle, and if the four initial circles have equal radii it is a round circle. Figure 3(c) is a plot of an Apollonian circle-packing: this is the limit set of a Kleinian group which also consists of even numbers of reflections in four initial circles, not drawn here (see Exercise 8, Section 9). For more examples, and beautiful pictures, see [21].

2. Dynamics of Rational Maps

2.1. *The Riemann sphere*

Let S^2 denote the unit sphere in $\mathbb{R}^3$, regard $\mathbb{C}$ as the plane $\mathbb{R}^2 \subset \mathbb{R}^3$ which cuts through S^2 at its equator, and let $N = (0,0,1)$ and $S = (0,0,-1)$ denote the north and south poles of S^2. Stereographic projection from N defines a homeomorphism π from the complement $S^2 \setminus \{N\}$ of N to $\mathbb{C}$. Extending π to send N to ∞ we obtain a bijection from S^2 to $\hat{\mathbb{C}} = \mathbb{C} \cup \infty$. We give S^2 the structure of a *Riemann surface* by equipping it with charts (homeomorphisms) $\phi_1 : \mathbb{C} \to S^2 \setminus \{N\}$ and $\phi_2 : \mathbb{C} \to S^2 \setminus \{S\}$ such that $\phi_2^{-1}\phi_1$ is an analytic bijection on the overlap. If we take ϕ_1 to be π^{-1}, and ϕ_2 to be the inverse of stereographic projection from S, followed by complex conjugation, the overlap $\phi_2^{-1}\phi_1$ is $z \to \bar{z}/|z|^2 = 1/z$.

Equally we can put a complex structure on $\hat{\mathbb{C}}$ by regarding it as the *complex projective line*

$$\mathbb{CP}^1 = \{\mathbb{C}^2 - (0,0)\}/\mathcal{R},$$

where $\mathcal{R}$ is the relation $(z,w) \sim (\lambda z, \lambda w)$ for $\lambda \in \mathbb{C} \setminus \{0\}$. An equivalence class $[z,w]$ contains $(z/w, 1)$ if $w \ne 0$ or $(1, w/z)$ if $z \ne 0$, so we may think of $\mathbb{CP}^1$ as the union of two copies of the complex plane glued together,

$\mathbb{C}_1 \cup \mathbb{C}_2/(z_1 \sim 1/z_2)$. The bijection $\mathbb{CP}^1 \leftrightarrow \hat{\mathbb{C}}$ is given by $[z,w] \leftrightarrow z/w$ when $w \neq 0$, and $[z,0] \leftrightarrow \infty$ for $z \neq 0$.

2.2. Basic essentials from complex analysis

An open connected set $\Omega \subset \mathbb{C}$ is called a *domain*.

A function $f : \Omega \to \mathbb{C}$ is said to be *differentiable* at $z_0 \in \Omega$ if the limit

$$f'(z_0) = \lim_{z \to z_0} \frac{f(z) - f(z_0)}{z - z_0} \quad \text{exists.}$$

If $f : \Omega \to \mathbb{C}$ is differentiable at all $z_0 \in \Omega$ it is said to be *holomorphic*.

Theorem 2.1. *Let f be holomorphic on the domain $\Omega \subset \mathbb{C}$. Let $z_0 \in \Omega$ and let R be the radius of the largest disc which has centre z_0 and is contained in Ω. Then for all z with $|z - z_0| < R$ the Taylor series $\sum_{n=0}^{\infty} a_n(z - z_0)^n$ for f at z_0 converges absolutely to the value $f(z)$.*

This is a classical theorem of complex analysis. A function expressible as a sum of a power series is called *analytic*: Theorem 2.1 says that a holomorphic function on a domain $\Omega \subset \mathbb{C}$ is analytic. The converse is also well known: the sum of a power series is holomorphic on the disc of convergence of the series, and its derivative is given by term-by-term differentiation.

There is a geometric interpretation for the statement that a function f is differentiable at z_0. If $f'(z_0) \neq 0$, then near z_0 we have

$$(f(z) - f(z_0)) \sim f'(z_0)(z - z_0)$$

so f acts on $(z - z_0)$ by multiplying it by the scale factor $|f'(z_0)|$ and turning it through an angle $\arg(f'(z_0))$. In particular if $f'(z_0) \neq 0$ the function f is *conformal* (angle-preserving) at z_0. If $f'(z_0) = 0$, then on a small disc centred at z_0 we have $f(z) \sim f(z_0) + a_n(z - z_0)^n$ for the first non-zero coefficient a_n, and f acts on this disc as an *n-to-1 branched covering map* (branched at z_0): note that if $n > 1$ then f is *not* conformal at z_0, indeed it multiplies angles at z_0 by n.

If $f : \Omega \to \mathbb{C}$ is holomorphic except at *isolated singularities* we say that f is *meromorphic* if all these singularities are either *removable* or *poles*, or equivalently if for each $z_0 \in \Omega$ there is a disc neighbourhood D of z_0 such that the Laurent series for f in the punctured disc $D \setminus \{z_0\}$ has the form $\sum_{n=-m}^{+\infty} a_n(z-z_0)^n$. Recall that z_0 is said to be a *pole of order m* if $m > 0$ is such that $a_{-m} \neq 0$ but $a_{-n} = 0$ for all $n > m$, and that z_0 is said to be *removable* if $a_{-n} = 0$ for all $n > 0$. When z_0 is a removable singularity we can set $f(z_0) = a_0$ and thereby extend f to a function differentiable at z_0, and when z_0 is a pole $\lim_{z \to z_0} f(z) = \infty$ so we can extend the definition of f by setting $f(z_0) = \infty$ and regard f as a continuous function $f : \Omega \to \hat{\mathbb{C}} = \mathbb{C} \cup \{\infty\}$. This extension is generally called *meromorphic* too.

There is a nice way to characterise a meromorphic function $f : \Omega \to \hat{\mathbb{C}}$ (where Ω is a domain in $\mathbb{C}$), making use of the 'duality' between '0' and '∞'. Let σ denote the function $z \to 1/z$. Then around any pole z_0 of f the function σf is analytic, since $f(z)$ has an expression as a Laurent series

$$f(z) = (z - z_0)^{-m} \sum_{n=0}^{\infty} b_n (z - z_0)^n \quad (b_0 \neq 0)$$

and taking the reciprocal of this we obtain an expression of the form:

$$\sigma f(z) = (z - z_0)^m \sum_{n=0}^{\infty} c_n (z - z_0)^n$$

where $c_0 = 1/b_0$. Thus $f : \Omega \to \hat{\mathbb{C}}$ is meromorphic if and only if f is analytic at points z_0 where $f(z_0) \neq \infty$ and σf is analytic at points z_0 where $f(z_0) \neq 0$.

Finally, we allow Ω to be a domain in $\hat{\mathbb{C}} = \mathbb{C} \cup \infty$, and we say that $f : \Omega \to \hat{\mathbb{C}}$ is *meromorphic at* ∞ if $f\sigma$ is meromorphic at 0. The functions $f : \hat{\mathbb{C}} \to \hat{\mathbb{C}}$ which are meromorphic on $\mathbb{C}$ and at ∞ are the *holomorphic* self-maps of $\hat{\mathbb{C}}$. They are the functions which, if we replace f by $\sigma f, f\sigma$ or $\sigma f \sigma$ as appropriate, have a Taylor series expansion at every point of $\hat{\mathbb{C}}$.

2.3. Rational maps and critical points

Theorem 2.2. *$f : \hat{\mathbb{C}} \to \hat{\mathbb{C}}$ is holomorphic if and only if f is a rational function, that is to say there exist polynomials $p(z), q(z)$, with complex coefficients, such that $f(z) = p(z)/q(z)$ for all $z \in \hat{\mathbb{C}}$.*

Proof. It is an elementary exercise to show that any rational map f is meromorphic both at points of $\mathbb{C}$ and at ∞, since by the Fundamental Theorem of Algebra f has the form

$$f(z) = c \frac{(z - \alpha_1)^{m_1} \cdots (z - \alpha_r)^{m_r}}{(z - \beta_1)^{n_1} \cdots (z - \beta_s)^{n_s}}.$$

For the converse, let $f : \hat{\mathbb{C}} \to \hat{\mathbb{C}}$ be holomorphic. Then f has finitely many poles (else $1/f$ has a convergent sequence of zeros, which, as a consequence of Theorem 2.1, is only possible if $1/f$ is identically zero). Let these poles be $\beta_1, \ldots, \beta_s$, of order $n_1, \ldots n_s$ respectively. Then

$$g(z) = (z - \beta_1)^{n_1} \cdots (z - \beta_s)^{n_s} f(z)$$

is analytic on $\mathbb{C}$, and so g can be written in the form

$$g(z) = \sum_{n=0}^{\infty} a_n z^n.$$

Since f is meromorphic at ∞ so is g. Thus $g\sigma$ is meromorphic at 0. So $\sum_{n=0}^{\infty} a_n z^{-n}$ has a pole or a removable singularity at $z = 0$. It follows that only finitely many of the a_n are non-zero and hence g is a polynomial. □

This is a powerful result: it tells us that any holomorphic $f : \hat{\mathbb{C}} \to \hat{\mathbb{C}}$ is determined by a *finite* set of data, for example the poles and zeros of f, with their multiplicities, together with the value of f at one other point.

Let $f(z) = p(z)/q(z)$, where p and q are polynomials of degree d_p and d_q respectively, with no common zeros. Then a generic point in $\hat{\mathbb{C}}$ has $\max(d_p, d_q)$ inverse images. We define the *degree* of f to be $\max(d_p, d_q)$.

A *critical point* of a rational map f is a point z_0 where the degree one term of the Taylor series for f vanishes, i.e., $f'(z_0) = 0$. As usual we replace f by $f\sigma$ here if $z_0 = \infty$, by σf if $f(z_0) = \infty$ and by $\sigma f \sigma$ if both are ∞, so that an appropriate Taylor series exists. Looked at topologically, a critical point of f is a *branch point* of f, a point z_0 such that $f(z) - f(z_0)$ has a factor $(z - z_0)^n$ for some $n > 1$, and thus in particular where $f^{-1}f(z_0)$ consists of less than $\deg(f)$ distinct points. Writing $f(z) = p(z)/q(z)$, we see that $f'(z) = 0 \Leftrightarrow q'(z)p(z) - p'(z)q(z) = 0$.

Proposition 2.3. *A degree d rational map has $2d - 2$ critical points (counted with multiplicity).*

Proof. In the generic case both p and q have degree d and $q'(z)p(z) - p'(z)q(z)$ is generically a polynomial of degree $2d - 2$ (since $q'(z)p(z)$ and $p'(z)q(z)$ have the same degree $2d - 1$ term). In the non-generic case we obtain the same result if we adopt the right notion of 'multiplicity' (see Proposition 3.5, in Section 3). □

2.4. Conformal automorphisms of $\hat{\mathbb{C}}$, $\mathbb{C}$ and $\mathbb{D}$

The invertible holomorphic maps $f : \hat{\mathbb{C}} \to \hat{\mathbb{C}}$ are the *conformal automorphisms* of the Riemann sphere. They form a group $\mathrm{Aut}(\hat{\mathbb{C}})$.

Proposition 2.4. $\mathrm{Aut}(\hat{\mathbb{C}})$ *consists of the rational maps of the form*

$$f(z) = \frac{az + b}{cz + d}$$

which have $a, b, c, d \in \mathbb{C}$ and $ad \neq bc$.

Proof. By Theorem 2.2, for f to be holomorphic it must be rational, but to be injective it must have degree 1. Conversely, any f of this form is invertible since it has inverse $f^{-1}(z) = (dz - b)/(-cz + a)$. □

Maps of the form $f(z) = (az+b)/(cz+d)$ having $a, b, c, d \in \mathbb{C}$ and $ad \neq bc$ are called *fractional linear* or *Möbius* transformations. We list some of their properties:

(1) Every invertible linear map $\alpha : \mathbb{C}^2 \to \mathbb{C}^2$ has the form

$$\begin{pmatrix} z \\ w \end{pmatrix} \to \begin{pmatrix} a & b \\ c & d \end{pmatrix} \begin{pmatrix} z \\ w \end{pmatrix} = \begin{pmatrix} az + bw \\ cw + dw \end{pmatrix}$$

and passes to a map $\mathbb{CP}^1 \to \mathbb{CP}^1$ which in our coordinate z/w on $\hat{\mathbb{C}}$ is

$$z/w \to \frac{az+bw}{cz+dw} = \frac{az/w+b}{cz/w+d}$$

(where $(a\infty + b)/(c\infty + d)$ is to be interpreted as a/c and so on).

(2) Composition of linear maps passes to composition of Möbius transformations. The group of all Möbius transformations is therefore

$$\mathrm{PGL}(2, \mathbb{C}) = \frac{\mathrm{GL}(2, \mathbb{C})}{\{\lambda I; \lambda \in \mathbb{C} \setminus \{0\}\}} = \frac{\mathrm{SL}(2, \mathbb{C})}{\{\pm I\}} = \mathrm{PSL}(2, \mathbb{C})$$

where $\mathrm{GL}(2, \mathbb{C})$ denotes the group of all invertible 2×2 matrices and $\mathrm{SL}(2, \mathbb{C})$ denotes those of determinant 1.

(3) Given any three distinct points $P, Q, R \in \hat{\mathbb{C}}$, there exists a unique Möbius transformation sending $P \to \infty, Q \to 0, R \to 1$, given by

$$\alpha(z) = \frac{(P-R)(Q-z)}{(Q-R)(P-z)}.$$

(Uniqueness follows from the easy exercise that the only Möbius transformation fixing $0, 1$ and ∞ is the identity.) It follows that given any other three distinct points $P', Q', R' \in \hat{\mathbb{C}}$ there exists a unique Möbius transformation sending $P \to P'$, $Q \to Q'$ and $R \to R'$.

(4) Given any four distinct points $P, Q, R, S \in \hat{\mathbb{C}}$, their *cross-ratio* is:

$$(P, Q; R, S) = \frac{(P-R)(Q-S)}{(Q-R)(P-S)} \in \hat{\mathbb{C}} \setminus \{0, 1, \infty\}.$$

It follows from (3) that $(P, Q; R, S) = \alpha(S)$, where α is the unique Möbius transformation sending $P \to \infty$, $Q \to 0$, $R \to 1$. Hence if γ is any Möbius transformation then $\alpha\gamma^{-1}$ is a Möbius transformation sending $\gamma(P) \to \infty$, $\gamma(Q) \to 0$ and $\gamma(R) \to 1$, and so $(\gamma(P), \gamma(Q); \gamma(R), \gamma(S)) = (P, Q; R, S)$.

Möbius transformations are conformal (since they are invertible and therefore have non-zero derivative everywhere). But conformality is a local property and we can prove a much stronger global property.

Proposition 2.5. *Möbius transformations send circles in $\hat{\mathbb{C}}$ to circles in $\hat{\mathbb{C}}$ (where a 'circle through ∞' is a straight line in $\mathbb{C}$).*

Proof. Every 'circle' in $\hat{\mathbb{C}}$ (including those through ∞) has the form
$$\alpha(x^2 + y^2) + 2\beta x + 2\gamma y + \delta = 0 \quad (\alpha, \beta, \gamma, \delta \in \mathbb{R})$$
in other words $Az\bar{z} + Bz + \bar{B}\bar{z} + C = 0$, where $A = \alpha \in \mathbb{R}$, $B = \beta - i\gamma \in \mathbb{C}$, and $C = \delta \in \mathbb{R}$. Let
$$z = \frac{aw + b}{cw + d}.$$
Now a direct substitution for z in the equation for a circle gives an equation of the same form for w, once the denominator has been cleared. $\square$

Corollary 2.6. *Any four distinct points $P, Q, R, S \in \hat{\mathbb{C}}$ lie on a common circle if and only if their cross-ratio $(P, Q; R, S)$ is real.*

Proof. Send P, Q, R to $\infty, 0, 1$ by a Möbius transformation. $\square$

Proposition 2.7. *The conformal automorphisms of $\mathbb{C}$ are the maps of the form $f(z) = az + b$, with $a, b \in \mathbb{C}$ and $a \neq 0$.*

Proof. Let f be a conformal automorphism of $\mathbb{C}$. Then $\lim_{z \to \infty} f(z) = \infty$ (this follows from the fact that f is a homeomorphism). Hence $\sigma f \sigma$ has a removable singularity at 0 and so f extends to a conformal automorphism of $\hat{\mathbb{C}}$. The result follows by Proposition 2.4. $\square$

We next identify the conformal automorphisms of the open unit disc $\mathbb{D} \subset \mathbb{C}$, using Schwarz's Lemma (which will be useful again later).

Lemma 2.8 (Schwarz's Lemma). *If f is holomorphic $\mathbb{D} \to \mathbb{D}$ and $f(0) = 0$ then $|f'(0)| \leq 1$. If $|f'(0)| = 1$ then $f(z) = \mu z$ for some $\mu \in \mathbb{C}$ with $|\mu| = 1$. If $|f'(0)| < 1$ then $|f(z)| < |z|$ for all $0 \neq z \in \mathbb{D}$.*

Proof. This is an application of the Maximum Modulus Principle. See any complex analysis textbook for details. $\square$

Proposition 2.9. *Every conformal automorphisms of $\mathbb{D}$ has the form*
$$f(z) = e^{i\theta} \frac{z - a}{1 - \bar{a}z}, \quad \theta \in \mathbb{R}, \ a \in \mathbb{D}.$$

Proof. Let f be a conformal automorphism of $\mathbb{D}$. Then $f^{-1}(0) = a \in \mathbb{D}$. The Möbius transformation
$$g(z) = \frac{z-a}{1-\bar{a}z}$$
sends a to 0 and the unit circle to itself, so it sends $\mathbb{D}$ to itself. Thus fg^{-1} is a conformal automorphism of $\mathbb{D}$ sending 0 to 0. From Schwarz's lemma it follows that $fg^{-1}(z) = \mu z$ for some μ with $|\mu| = 1$. $\square$

Corollary 2.10. *The conformal automorphisms of the upper half-plane $\mathcal{H}_+$ are the Möbius transformations*
$$f(z) = \frac{az+b}{cz+d}$$
having $a, b, c, d \in \mathbb{R}$ and $ad \neq bc$.

Proof. Take any Möbius transform M which sends the upper half-plane $\mathcal{H}_+$ bijectively onto $\mathbb{D}$ (exercise: write one down). The conformal automorphisms of $\mathcal{H}_+$ are the maps $M^{-1}gM$ where g runs through the conformal automorphisms of $\mathbb{D}$ given by Proposition 2.9 (details: exercise). $\square$

We recall that not only is there is a conformal bijection between $\mathcal{H}_+$ and $\mathbb{D}$, but that the Riemann Mapping Theorem states that for every simply-connected domain $U \subset \mathbb{C}$ ($U \neq \mathbb{C}$) there is a conformal bijection between U and $\mathbb{D}$. An important generalisation of this that we shall use several times, explicitly or implicitly, but which we will not prove here, is the following (proved by Poincaré and Koebe).

Theorem 2.11 (The Uniformisation Theorem). *Every simply-connected Riemann surface is conformally bijective to one of $\hat{\mathbb{C}}$, $\mathbb{C}$ or $\mathbb{D}$.*

2.5. The Poincaré metric on the upper half-plane

Define the *infinitesimal Poincaré metric* on the upper half-plane by
$$ds = \frac{\sqrt{(dx)^2 + (dy)^2}}{y}.$$

Proposition 2.12. *ds is invariant under $\mathrm{PSL}(2, \mathbb{R})$.*

Proof. Every element of $\mathrm{PSL}(2, \mathbb{R})$ can be written as a composition of transformations of the type $z \to z + \lambda$ ($\lambda \in \mathbb{R}$), $z \to \mu z$ ($\mu \in \mathbb{R}^{>0}$) and $z \to -1/z$, and it is easily checked that each preserves ds. $\square$

A path of shortest length between P and Q in $\mathcal{H}_+$ is called a *geodesic*.

Proposition 2.13. *Between any two distinct points P and Q in $\mathcal{H}_+$ there is a unique geodesic. It is the segment between P and Q of the (unique) euclidean semicircle through P and Q which meets $\hat{\mathbb{R}}$ orthogonally. The distance between P and Q (in the Poincaré metric) is $\ln(|(P,Q;A,B)|)$ where A and B are the points where the semicircle meets $\hat{\mathbb{R}}$.*

Proof. If P and Q are on the imaginary axis, the straight line path γ_1 from P to Q is shorter than any other path γ_2 from P to Q, since

$$\int_{\gamma_2} \frac{1}{y}\sqrt{(dx)^2 + (dy)^2} > \int_{\gamma_2} \frac{1}{y} dy = \int_{\gamma_2} \frac{1}{y} dy.$$

For $P = i$ and $Q = it$ (real $t > 1$) the hyperbolic distance from P to Q is

$$\int_1^t \frac{1}{y} dy = \ln t = \ln|(i, it; 0, \infty)|.$$

The result follows, since given any P', Q' in $\mathcal{H}_+$ there is an element of $\mathrm{PSL}(2,\mathbb{R})$ which sends P to P' and Q to Q', and moreover this Möbius transformation sends the positive imaginary axis to a semicircle with ends on the extended real axis $\hat{\mathbb{R}}$ and preserves cross-ratios. □

We can transfer the Poincaré metric to $\mathbb{D}$, using any Möbius transformation M sending $\mathcal{H}_+ \to \mathbb{D}$. On $\mathbb{D}$ the infinitesimal metric becomes $2|dz|/(1-|z|^2)$, the geodesics are the arcs of circles that meet the boundary $\partial \mathbb{D}$ orthogonally, and the distance between any P and Q is $\ln(|(P,Q;A,B)|)$ where A and B are the points where the geodesic through P and Q meets $\partial \mathbb{D}$ (exercise).

Corollary 2.14. *The group of conformal automorphisms of the upper half-plane, $\mathrm{PSL}(2,\mathbb{R})$, is also the group of orientation-preserving isometries of the upper half-plane (equipped with the Poincaré metric).*

Proof. Every element of $\mathrm{PSL}(2,\mathbb{R})$ preserves the Poincaré metric since it preserves the upper half-plane, the real axis and cross-ratios. For the converse, first observe that to prove that an isometry α is conformal at some particular point $z_0 \in \mathcal{H}_+$, we can reduce to the case that $\alpha(z_0) = z_0$ by composing by a suitable element of $\mathrm{PSL}(2,\mathbb{R})$ (which we know to be an isometry by Proposition 2.12). But any such α is conformally conjugate (via a Möbius transformation) to an isometry β of $\mathbb{D}$ (equipped with the Poincaré metric) such that $\beta(0) = 0$, and an isometry β with this property is necessarily a rigid rotation of $\mathbb{D}$ since it maps each circle centred at 0 to itself by a rigid rotation through the same angle. □

2.6. Conjugacies, fixed points and multipliers

Rational maps f, g are said to be *conjugate* if there exists a Möbius transformation h such that $g = hfh^{-1}$, in other words such that the following diagram commutes:

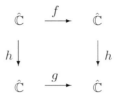

Conjugate maps have identical dynamical behaviour (think of h as a 'change of coordinate system'). We can often put a rational map into a simpler form by applying a suitable conjugacy.

Examples

(1) A rational map f is conjugate to a polynomial if and only if there exists a point $z_0 \in \hat{\mathbb{C}}$ such that $f^{-1}(z_0) = \{z_0\}$. (**Proof.** Use as a conjugacy h any Möbius transformation which sends z_0 to ∞. Details: exercise.)
(2) A rational map f is conjugate to a polynomial of the form $z \to z^n$ (some $n > 0$) if and only if there exist distinct points $z_0, z_1 \in \hat{\mathbb{C}}$ such that $f^{-1}(z_0) = \{z_0\}$ and $f^{-1}(z_1) = \{z_1\}$. (**Proof.** Exercise.)
(3) Every degree 2 polynomial $z \to \alpha z^2 + \beta z + \gamma$ ($\alpha \neq 0$) is conjugate to a unique one of the form $z \to z^2 + c$. (**Proof.** Exercise 3, Section 9).

A *fixed point* of a rational map f is a point $z_0 \in \hat{\mathbb{C}}$ such that $f(z_0) = z_0$. The *multiplier* of f at z_0 is the derivative $f'(z_0) = \lambda$ (for $z_0 = \infty$ the multiplier is $g'(0)$, where $g = \sigma f \sigma$). We say that z_0 is:

attracting if $|\lambda| < 1$ (if $\lambda = 0$ we say z_0 is *superattracting*);
repelling if $|\lambda| > 1$;
neutral if $|\lambda| = 1$, i.e., $\lambda = e^{2\pi i \theta}$ for some $\theta \in \mathbb{R}$.

As we shall see in Section 4.1, the dynamical behaviour around a neutral periodic point depends on whether θ is rational or irrational, and the irrational case can be further subdivided into 'linearisable' and 'non-linearisable'.

Proposition 2.15. *When the function f is conjugated by a Möbius transformation h, any fixed point z_0 of f is sent to a fixed point $w_0 = h(z_0)$ of $g = hfh^{-1}$, and the multiplier of the fixed point w_0 for g is equal to the multiplier of the fixed point z_0 for f.*

Proof. If z_0 is a fixed point of f and $w_0 = h(z_0)$ then
$$g(w_0) = gh(z_0) = hf(z_0) = w_o$$
and, by the chain rule for differentiation,
$$g'(w_0) = h'(w_0)f'(z_0)(h^{-1})'(w_0).$$
But since h is differentiable, has differentiable inverse and sends z_0 to w_0,
$$(h^{-1})'(w_0) = \frac{1}{h'(z_0)}$$
and hence $g'(w_0) = f'(z_0)$. □

Proposition 2.15 says that a conjugacy sends a fixed point of f to a fixed point of g having the same dynamical behaviour (attractor, repellor etc.).

A point z_0 is said to *periodic* of period n for f if $f^n(z_0) = z_0$ but $f^j(z_0) \neq z_0$ for $0 < j < n$. The *multiplier* of the periodic orbit $\{z_0, f(z_0) = z_1, f(z_1) = z_2, \ldots, f(z_{n-1}) = z_0\}$ is defined to be $(f^n)'(z_0)$. Note that $(f^n)'(z_0) = f'(z_0)f'(z_1)\ldots f'(z_{n-1})$ by the chain rule.

Proposition 2.16. *When the function f is conjugated by a Möbius transformation h, any orbit of period n of f is sent to an orbit of period n of $g = hfh^{-1}$, and the two orbits have the same multiplier.*

Proof. Exercise. □

3. The Fatou and Julia Sets of a Rational Map

3.1. *Equicontinuity and the Fatou and Julia sets*

We define the *spherical metric* on the unit sphere S^2 by setting the distance between two points to be the shortest length of a great circle path between them. On the Riemann sphere, parameterised as the extended complex plane $\mathbb{C} \cup \infty$, the infinitesimal spherical metric is:
$$ds = \frac{2|dz|}{1+|z|^2}.$$

Warning. The spherical metric is not preserved by $\mathrm{Aut}(\hat{\mathbb{C}})$, but conjugating by any particular conformal automorphism sends the spherical metric to a Lipschitz equivalent metric, since $\hat{\mathbb{C}}$ is compact.

Let f be a rational map and z_0 be a point of $\hat{\mathbb{C}}$. We say that the family of iterates $\{f^n\}_{n\geq 0}$ is *equicontinuous* at z_0 if given any $\epsilon > 0$ there exists $\delta > 0$ such that for all $n \geq 0$, $d(f^n(z), f^n(z_0)) < \epsilon$ whenever $d(z, z_0) < \delta$. (Here d is the spherical metric on $\hat{\mathbb{C}}$.) Think of this as saying that *'any orbit which that starts near z_0 remains close to the orbit of z_0 for all time'*.

The *Fatou set* $F(f)$ of f is defined to be the largest open subset of $\hat{\mathbb{C}}$ on which the family $\{f^n\}_{n\geq 0}$ is equicontinuous at every point. The *Julia set* $J(f)$ of f is defined as its complement $\hat{\mathbb{C}} \setminus F(f)$, and can be thought of as where the dynamics exhibits *'sensitive dependence on initial conditions'*.

The following properties follow at once from our definitions above, and the fact that the image of any open set under a rational map is itself open:

(1) $F(f)$ is open; hence $J(f)$ is closed and therefore compact.
(2) $F(f)$ is *completely invariant*, that is $f(F(f)) = F(f) = f^{-1}(F(f))$; hence $J(f)$ is also completely invariant.

3.2. *Equicontinuity and normality*

Which families $\mathcal{F}$ of analytic maps $f : \Omega \to \hat{\mathbb{C}}$ are equicontinuous? Our first step towards an answer is to interpret Schwarz's Lemma in the language of hyperbolic geometry:

Proposition 3.1. *If f is a holomorphic map $\mathbb{D} \to \mathbb{D}$ then f is non-increasing in the Poincaré metric.*

Proof. Let z_0, z_1 be any two points in $\mathbb{D}$. Let $f(z_0) = w_0$ and $f(z_1) = w_1$. Choose isometries h, k of the Poincaré disc $\mathbb{D}$ (Möbius transformations) such that $h(0) = z_0$ and $k(0) = w_0$. Let $z_1' = h^{-1} z_1$ and $w_1' = k^{-1} w_1$. Now $k^{-1} f h$ is a holomorphic map of $\mathbb{D}$ to itself sending 0 to 0 and z_1' to w_1'. Hence $|w_1'| \leq |z_1'|$ by Schwarz's lemma, and so $d(0, w_1') < d(0, z_1)$ in the Poincaré metric. But $d(w_0, w_1) = d(0, w_1')$ and $d(z_0, z_1) = d(0, z_1')$ (as h and k are isometries). □

Corollary 3.2. *Every family of holomorphic maps $\mathbb{D} \to \mathbb{D}$ is equicontinuous with respect to the spherical metric on $\mathbb{D} \subset \hat{\mathbb{C}}$.*

Proof. It follows at once from Proposition 3.1 that every such family $\mathcal{F}$ is equicontinuous with respect to the Poincaré metric on $\mathbb{D}$. However given any point z_0 we can find a small disc around z_0 and a constant k such that the distance between any two points z, z' in this disc in the Poincaré metric is less than k times the distance in the spherical metric. Equicontinuity at z_0 follows, since the spherical distance between the images $f(z), f(z')$ of two points under $f \in \mathcal{F}$ is less than or equal to the Poincaré distance between these images, this being true for any pair of points in $\mathbb{D}$. $\square$

Example. By Corollary 3.2, the family $\{z \to z^{2^n}\}_{n \geq 0}$ is equicontinuous on $\mathbb{D}$: thus the Fatou set of $z \to z^2$ contains $\{z : |z| < 1\}$. Conjugating by $\sigma : z \to 1/z$ we see that the Fatou set of $z \to z^2$ also contains $\{z : |z| > 1\}$. Since no point on the unit circle is in the Fatou set (recall that $z \to z^2$ doubles distance there), we now have a proof that the Fatou and Julia sets of $z \to z^2$ are as claimed in Section 1.2.

Note that it follows from Corollary 3.2 that every *bounded* family of holomorphic maps $\mathbb{D} \to \mathbb{C}$ is equicontinuous.

There are two approaches to defining the Fatou set of a rational map f: either as the *equicontinuity set* of the family of iterates of f, or as the *normality set* of this family. Let Ω be a domain in $\hat{\mathbb{C}}$. A family $\mathcal{F}$ of maps $\Omega \to \hat{\mathbb{C}}$ is called *normal* if every sequence in $\mathcal{F}$ contains a subsequence which converges *locally uniformly* to a map $f : \Omega \to \hat{\mathbb{C}}$ (not necessarily in $\mathcal{F}$).

Example. $\{z \to z^{2^n}\}_{n \geq 0}$ is a normal family on $\mathbb{D}$, since these maps converge locally uniformly on $\mathbb{D}$ to the constant map $z \to 0$.

Theorem 3.3 (Arzelà–Ascoli). *Let Ω be a domain in $\hat{\mathbb{C}}$. A family of continuous maps $\Omega \to \hat{\mathbb{C}}$ is normal if and only if it is equicontinuous at all $z \in \Omega$.*

Proof. See for example Ahlfors' book 'Complex Analysis' [1]. $\square$

We remind the reader that we use the *spherical metric* on $\hat{\mathbb{C}}$, in the definitions of both *equicontinuity* and *local uniform convergence* (used in defining the notion of a *normal family*). It is more elegant mathematically to develop the whole Fatou–Julia theory via normality rather than equicontinuity but the latter is perhaps easier to comprehend dynamically.

This brings us to a theorem central to Fatou–Julia theory.

Theorem 3.4 (Montel). *Let Ω be a domain in $\hat{\mathbb{C}}$. Every family of analytic maps $\Omega \to \hat{\mathbb{C}} \setminus \{0, 1, \infty\}$ is normal (or equivalently, by Arzelà–Ascoli, equicontinuous).*

Proof. This is a consequence of the fact that the *universal cover* of $\hat{\mathbb{C}} \setminus \{0, 1, \infty\}$ is $\mathbb{D}$. Thus on any simply-connected neighbourhood $U \subset \Omega$ we can lift our family of analytic maps to a family of analytic maps $U \to \mathbb{D}$. The result now follows from the fact that this lifted family is normal (by Corollary 3.2 and Theorem 3.3). For details, see [4] or [19]. □

We can replace the points $0, 1$ and ∞ in the statement of Montel's theorem by any other three points of $\hat{\mathbb{C}}$ (just compose with a suitable Möbius transformation). Montel's theorem is a much more powerful result than our earlier observation that any family of maps with a common bound is equicontinuous. One should compare it with Picard's theorem that any holomorphic function $\mathbb{C} \to \hat{\mathbb{C}} \setminus \{0, 1, \infty\}$ is constant, which in turn is much more powerful than Liouville's theorem that a bounded holomorphic function on $\mathbb{C}$ is constant.

3.3. Counting critical points, and the exceptional set

Before considering the many properties of Julia sets which follow from Montel's theorem, we make a brief excursion into topology to count critical points and derive some consequences for *finite* completely invariant sets.

The *valency* of a critical point c of a rational map f is ν_c, where locally near c the map f has the form $z \to kz^{\nu_c}$ (plus higher order terms). In other words the valency is the 'degree of branching' at c.

Proposition 3.5 (Riemann–Hurwitz Formula). *If f is a rational map of degree d, then $\sum_c (\nu_c - 1) = 2d - 2$, where the sum is over all critical points c of f.*

Proof. Triangulate the target copy of $\hat{\mathbb{C}}$ in such a way that the critical values of f are all vertices, and pull this triangulation back, via f, to a triangulation of the source copy of $\hat{\mathbb{C}}$. The Euler characteristic of $\hat{\mathbb{C}}$ (number of triangles minus number of edges plus number of vertices) is 2. Apart from at critical points, f is d to 1, and so $2d - \sum_c(\nu_c - 1) = 2$. □

Corollary 3.6. *Let f be a rational map with $\deg(f) \geq 2$, and E be a finite completely invariant subset of $\hat{\mathbb{C}}$. Then E contains at most 2 points.*

Proof. Suppose E contains k points. Then f must permute these points (since every surjection of a finite set to itself is a bijection) and hence for some q the iterate $f^q = g$ is the identity on E. Suppose g has degree d. Each point $z \in E$ must be a critical point of g, of valency d, else $g^{-1}(z)$ would contain points other than z. Hence $k(d-1) \leq 2d - 2$ by Proposition 3.5, and therefore $k \leq 2$. □

The *exceptional set* $E(f)$ of a rational map is defined to be the union of all finite completely invariant sets. Corollary 3.6 says that $|E(f)| \leq 2$. Note that if $|E(f)| = 1$ then f is conjugate to a polynomial (just conjugate by a Möbius transformation sending the exceptional point to ∞), and if $|E(f)| = 2$ then f is conjugate to some $z \to z^d$, with d a positive or negative integer (just send the two exceptional points to ∞ and 0).

3.4. Properties of Julia sets

The Julia set $J(f)$ of every rational map f of degree ≥ 2 satisfies:

(1) $J(f) \neq \emptyset$.
(2) $J(f)$ is infinite.
(3) $J(f)$ is contained in every completely invariant closed subset of $\hat{\mathbb{C}}$ which contains at least three points.
(4) $J(f)$ is perfect, i.e., every $z \in J(f)$ is an accumulation point of $J(f)$ (so in particular $J(f)$ is uncountable).
(5) $J(f)$ is either the whole of $\hat{\mathbb{C}}$ or it has empty interior.

We omit proofs (for these see, for example, [4] or [19]) but remark that once we have proved property (1) the others follow as consequences of Montel's Theorem and Corollary 3.6. We also remark that there exist rational maps f having $J(f) = \hat{\mathbb{C}}$ (Lattès (1918) gave the example: $z \to (z^2+1)^2/4z(z^2-1)$) but that for a *polynomial* map the Fatou set always contains the point ∞ and hence is non-empty.

3.5. An algorithm for plotting $J(f)$

Proposition 3.7. *If* $\deg(f) \geq 2$ *and* U *is any open set in* $\hat{\mathbb{C}}$ *meeting* $J(f)$, *then* $\bigcup_{n=0}^{\infty} f^n(U) \supset \hat{\mathbb{C}} \setminus E(f)$.

Proof. If $\bigcup_{n=0}^{\infty} f^n(U)$ misses three or more points of $\hat{\mathbb{C}}$ then $\{f^n\}_{n \geq 1}$ are a normal family on U by Montel, contradicting $U \cap J \neq \emptyset$. But if a non-exceptional z lies in $\hat{\mathbb{C}} \setminus \bigcup_{n=0}^{\infty} f^n(U)$ then for some m and n a point of $f^{-m}(z)$ must lie in $f^n(U)$ (since $\bigcup_{m \geq 0} f^{-m}(z)$ is infinite). Hence $z \in f^{m+n}(U)$, which is a contradiction. □

This tells us, for example, that a small piece of $J(f)$ around a repelling fixed point is expanded by f^n to cover the whole of $J(f)$ when n is large: so $J(f)$ exhibits self-similarity on different scales.

Corollary 3.8. *If z_0 is not in $E(f)$, then $J(f) \subset \overline{\bigcup_{n \geq 0} f^{-n}(z_0)}$.*

Proof. Take any $z \in J(f)$ and neighbourhood U of z. By Proposition 3.7 the given point z_0 lies in some $f^n(U)$. Hence $f^{-n}(z_0) \cap U \neq \emptyset$. □

We deduce a simple algorithm for plotting $J(f)$. Start at any (non-exceptional) point z_0 and plot all its images under f^{-1}, then all of their images under f^{-1} etc., or alternatively plot $z_0, z_1, z_2, \ldots$ where each z_{j+1} is a random choice from the (finite) set of values of $f^{-1}(z_j)$. The resulting set accumulates on the whole of $J(f)$. Indeed, if one starts at a point z_0 known to be in $J(f)$ (e.g., a repelling fixed point) then $J(f) = \overline{\bigcup_{n \geq 0} f^{-n}(z_0)}$, so all points plotted are in the Julia set, not just accumulating on it.

3.6. Julia sets and repelling periodic points

From the definition of the Julia set $J(f)$, every repelling periodic point of f lies in $J(f)$. However it is also true that every point of $J(f)$ has a periodic point arbitrarily close to it.

Theorem 3.9. *If $\deg(f) \geq 2$ then $J(f)$ is contained in the closure of the set of all periodic points of f.*

Proof. Let $z_0 \in J(f)$ and assume z_0 is not a critical value of f (without loss of generality, since there are only a finite set of critical values and $J(f)$ is perfect). Then z_0 has a neighbourhood U on which two distinct branches of f^{-1} are defined. Denote these by $h_1 : U \to U_1$ and $h_2 : U \to U_2$ (where the sets U_1 and U_2 are disjoint). Suppose (for a contradiction) that U contains no periodic point of f. For each $z \in U$ set

$$g_n(z) = \frac{(f^n(z) - h_1(z))(z - h_2(z))}{(f^n(z) - h_2(z))(z - h_1(z))}.$$

Then for each value of n, $g_n(z) \neq 0, 1, \infty$ for $z \in U$ (else $\exists z \in U$ with $f^n(z) = h_1(z)$, $f^n(z) = z$ or $f^n(z) = h_2(z)$ and in each case f would have a periodic point). So by Montel's theorem the $\{g_n\}$ form a normal family. It follows (by an elementary exercise in analysis) that the $\{f^n\}$ are a normal family, contradicting the hypothesis that $z_0 \in J(f)$. □

In fact $J(f)$ is the closure of the set of all *repelling* periodic points of f: this follows from Theorem 3.9 together with the observation that every repelling periodic point of f lies in $J(f)$, and the theorem (of Fatou) that there are only finitely many non-repelling periodic orbits (see [4] or [19]). Thus the observation in Section 1.2 that the Julia set of $z \to z^2$ is the closure of the set of repelling periodic points is true for all rational functions.

3.7. The Julia set of $q_c : z \to z^2 + c$ when $q_c^n(0)$ tends to ∞

A *Cantor set* is a topological space homeomorphic to the space $C = \{0,1\}^{\mathbb{N}}$ of all infinite sequences of 0's and 1's, equipped with the product topology. Recall that every perfect totally disconnected compact subset of $\mathbb{R}^n$ is homeomorphic to C, so we can speak of *the* Cantor set.

Proposition 3.10. *When $q_c^n(0)$ tends to ∞ as n tends to ∞, the Julia set $J(q_c)$ is homeomorphic to the Cantor set C, and the action of q_c on $J(q_c)$ is conjugate to that of the shift on C.*

Proof. See [4] or [19]. We outline the argument first in the case that $|c| > 2$ (it is easy to prove that $|c| > 2 \Rightarrow \lim_{n\to\infty} q_c^n(0) = \infty$). Let γ_0 be the circle which has centre 0 and radius $|c|$, so passes through the critical value c. It is easily shown that if $|c| > 2$ then $|q_c(z)| > |z|$ whenever $|z| \geq |c|$. It follows that the curve $\gamma_1 = q_c^{-1}(\gamma_0)$ (which is a lemniscate with 'crossing point' the critical point 0) lies inside γ_0, and $q_c^{-1}(\gamma_1)$ consists of a lemniscate inside each lobe of γ_1, and so on (Fig. 4).

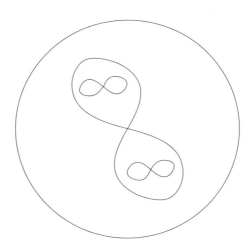

Fig. 4.   The circle γ_0 and its inverse images (Proposition 3.10).

Let D be any closed disc containing γ_1 and contained in γ_0. Label the two discs making up $q_c^{-1}(D)$ as D_0 and D_1. Continue inductively, labelling the 2^{n+1} components of $q_c^{-(n+1)}(D)$ by

$$D_{0s} = D_0 \cap q_c^{-1}(D_s), \quad D_{1s} = D_1 \cap q_c^{-1}(D_s)$$

for any length n sequence s of 0's and 1's. Set

$$\Lambda = \bigcap_{n=1}^{\infty} q_c^{-n}(D).$$

One can show that Λ consists of points D_s, each labelled by an infinite sequence s of 0's and 1's, and the action of q_c on Λ is conjugate to the action of the shift on these sequences. Moreover Λ is a minimal closed completely invariant set and is therefore the Julia set $J(q_c)$.

In the general case that $q_c^n(0) \to \infty$, the Böttcher conjugacy on a neighbourhood of ∞ (Theorem 4.4 in the next section) provides a simple closed Jordan curve through the critical value c that one can use in place of γ_0 in the argument above. $\square$

4. Fatou Components and Linearisation Theorems

The *basin* of an attracting fixed point z_0 is $\{z : \lim_{n\to\infty} f^n(z) = z_0\}$; the *immediate basin* is the component of this set containing z_0. There are similar definitions for an attracting n-cycle: here the immediate basin consists of those components of the basin which contain points of the cycle.

Theorem 4.1. *For every rational map f of degree at least two, the immediate basin of an attractive cycle contains a critical point.*

Proof. Without loss of generality we suppose z_0 to be an attracting *fixed point*. If z_0 is superattracting, the result is obvious. If z_0 is attracting but not superattracting, then there is a neighbourhood U of z_0 such that $f(U) \subset U$ and $f|_U$ is injective. Let $V = f(U)$ and consider the branch of f^{-1} sending V to U. If f has no *critical value* in U, this branch can be extended to the whole of U and hence f^{-2} has a well-defined branch on V. Repeat. If some $f^{-n}(V)$ contains a critical value, then the basin contains a critical point. But if not, then $\{f^{-n}\}_{n>0}$ are all defined on V and have images in the immediate basin. But then they are an equicontinuous family (by Montel): this is impossible as z_0 is a *repelling* fixed point for f^{-1}. $\square$

Corollary 4.2. *If f has degree d, then it has at most $2d - 2$ attracting or superattracting cycles.*

4.1. Linearisation theorems

A periodic point of f is a fixed point of some f^n, so it will be enough for us to consider a fixed point of f. Around such a point, which without loss of generality we take to be 0, f has Taylor series $f(z) = \lambda z + O(z^2)$. If there is a neighbourhood U of 0 on which f is *analytically conjugate* to $z \to \lambda z$, we say that f is *linearisable*.

Theorem 4.3 (Koenigs' Linearisation Theorem 1884). *If $f(z) = \lambda z + O(z^2)$ with $\lambda \neq 0$ and $|\lambda| \neq 1$, then f is linearisable on a neighbourhood of 0.*

Proof. Assume first that $0 < |\lambda| < 1$. Set
$$h_n(z) = \frac{1}{\lambda^n} f^n(z).$$
Then, by construction $h_n f(z) = \lambda h_{n+1}(z)$, and it suffices to show that the $\{h_n\}$ converge locally uniformly to a function h, since then $hf = \lambda h$. See the example below for a sketch proof in the case of a particular example, and Milnor [19, Theorem 8.2] for the general case. For $1 < |\lambda| < \infty$ one can proceed in exactly the same fashion, but with f^{-1} in place of f. □

Example. $f(z) = \lambda z + z^2$ (where $|\lambda| < 1$). Here the orbit of any initial point z_0 is:
$z_1 = f(z_0) = \lambda z_0 (1 + z_0/\lambda)$
$z_2 = f(z_1) = \lambda z_1 (1 + z_1/\lambda) = \lambda^2 z_0 (1 + z_0/\lambda)(1 + z_1/\lambda)$
and so on.

Thus $h_n(z_0) = z_0 (1 + z_0/\lambda)(1 + z_1/\lambda) \cdots (1 + z_{n-1}/\lambda)$ where $\{z_n\}$ is the orbit of z_0. As n goes to infinity, z_n goes to 0, and the family $\{h_n\}$ converges locally uniformly to
$$h(z_0) = z_0 \prod_0^\infty \left(1 + \frac{z_n}{\lambda}\right).$$

Theorem 4.4 (Böttcher 1904). *If $f(z) = z^k + O(z^{k+1})$ ($k \geq 2$ integer), then f is conjugate to $z \to z^k$ on a neighbourhood of 0.*

Proof. Analogously to the proof of Theorem 4.3, we set $h_n(z) = (f^n(z))^{1/k^n}$. Then $h_n f(z) = (h_{n+1}(z))^k$ and the $\{h_n\}$ converge locally uniformly to a function h conjugating f to $z \to z^k$. □

The proof above is only a sketch. See [19, Theorem 9.1], for details. The right choice of branch of k^nth root in the definition of h_n is important, but rather than fill in the details in general, we consider a particular example that will be useful later.

Example. Consider $f : z \to z^2 + c$ near the fixed point ∞. Write this map as $z \to z^2(1 + c/z^2)$.

$$z_1 = f(z_0) = z_0^2(1 + c/z_0^2)$$
$$z_2 = f(z_1) = z_1^2(1 + c/z_1^2) = z_0^4(1 + c/z_0^2)^2(1 + c/z_1^2)$$

and so on.

Thus $h_n(z_0) = z_0(1+c/z_0^2)^{1/2}(1+c/z_1^2)^{1/4} \ldots (1+c/z_{n-1}^2)^{1/2^n}$ where the choice of each root is the obvious one coming from the binomial expansion. As n tends to ∞ the z_n tend to ∞ (since z_0 is outside the filled Julia set). So the h_n converge (locally uniformly) to

$$h(z_0) = z_0 \prod_0^\infty \left(1 + \frac{c}{z_n^2}\right)^{1/2^{n+1}}.$$

What can be said about linearisability near a *neutral* fixed point? Suppose $f(z) = \lambda z + O(z^2)$, with $|\lambda| = 1$.

Case 1: $\lambda = e^{2\pi i p/q}$ (in this case $z = 0$ is called a *parabolic* fixed point).

Example. $f(z) = z + z^{n+1}$.

See Fig. 5: the 'attracting petals' bounded by dashed lines are mapped into themselves and each initial point in a petal has orbit which eventually converges to the fixed point along a direction tangent to the mid-line of the petal. The Julia set (not marked) heads off from the fixed point in directions tangent to the repelling axes (between the petals).

Fig. 5. Dynamics of $z \to z + z^{n+1}$ for $n = 1$, $n = 2$ and $n = 3$.

A rational map f is not linearisable around a parabolic fixed point (unless $f(z) = \lambda z$), since $f^q \neq identity$. But by analysing the local power series expansion of $f(z)$ it can be shown that the parabolic point itself lies in the Julia set and its the basin of attraction lies in the Fatou set (see [19, Lemma 10.5]).

The dynamical behaviour around a parabolic fixed point (or cycle) has a very particular form, that of a *Leau–Fatou flower*, with 'attracting petals' contained within the Fatou set, as illustrated in the examples above. For $\lambda = e^{2\pi i p/q}$ this flower has kq petals, where $k \geq 1$ (see, for example, [19, Theorem 10.7]). The study of holomorphic germs around parabolic points and cycles contains deep and interesting results: Chapter 10 of Milnor's book is an excellent starting point to learn more about this.

Case 2: $\lambda = e^{2\pi i \alpha}$ with α irrational.

Write α as a continued fraction

$$\alpha = a_0 + \cfrac{1}{a_1 + \cfrac{1}{a_2 + \cdots}} = [a_0, a_1, a_2, \ldots]$$

and let p_n/q_n (in lowest terms) be the value of its nth truncation $[a_0, a_1, \ldots, a_n]$. For example the golden mean $[0, 1, 1, 1, 1, \ldots]$ has $p_1/q_1 = 1/1, p_2/q_2 = 1/2, p_3/q_3 = 2/3, p_4/q_4 = 3/5, p_5/q_5 = 5/8$ and so on.

We say that α satisfies the *Brjuno condition* if and only if

$$\sum_1^\infty \frac{\log(q_{n+1})}{q_n} < \infty.$$

Write $\mathcal{B}$ for the set of real numbers satisfying the Brjuno condition.

Theorem 4.5 (Brjuno, 1965). $\alpha \in \mathcal{B} \Rightarrow$ all $z \to e^{2\pi i \alpha} z + O(z^2)$ are linearisable.

Theorem 4.6 (Yoccoz, 1988). $\alpha \notin \mathcal{B} \Rightarrow z \to e^{2\pi i \alpha} z + z^2$ is not linearisable. (See [26].)

When a linearisation exists its domain (contained in the Fatou set) is known as a *Siegel disc*. The irrational neutral points which are not linearisable lie in $J(f)$ and are known as *Cremer points*.

4.2. The classification of types of Fatou component

In 1985 Sullivan proved the following, which had been conjectured by Fatou.

Theorem 4.7 ('No Wandering Domains'). *For f rational, every component of $F(f)$ is either periodic or preperiodic.*

Proof. See [19, Appendix F]. The basic idea is that if there were a wandering domain then using the Measurable Riemann Mapping Theorem (see Section 7) one could construct an infinite-dimensional family of perturbations of f, all of them rational and topologically conjugate to f, but this is impossible since a rational map f is determined by a finite set of data. □

A consequence of this theorem (see [19, Chapter 16]) is that for a polynomial the only possible components of the Fatou set are components of the basin of a superattracting periodic cycle, or an attracting periodic cycle, or a rational neutral periodic orbit, or a periodic cycle of Siegel discs. There is one other type that can occur for rational, but not polynomial, f: components of the basin of a periodic cycle of *Herman rings* (annuli carrying dynamics conjugate to an irrational rotation).

5. Hyperbolic 3-space and Kleinian Groups

Let $\mathcal{H}^3_+ = \{(x_1, x_2, x_3) \in \mathbb{R}^3 : x_3 > 0\}$. Define the infinitesimal metric:
$$ds = \frac{1}{x_3}\sqrt{(dx_1)^2 + (dx_2)^2 + (dx_3)^2}.$$
With this metric $\mathcal{H}^3_+$ becomes the *upper half-space model of hyperbolic 3-space*. The geodesics are the semicircles in $\mathcal{H}^3_+$ orthogonal to the plane $x_3 = 0$. Now regard the plane $x_3 = 0$ as the complex plane $\mathbb{C}$ (via $(x_1, x_2, 0) \leftrightarrow z = x_1 + ix_2$), add the point '$\infty$', and think of $\hat{\mathbb{C}}$ as the *boundary* of $\mathcal{H}^3_+$. Every fractional linear map
$$\alpha : z \to \frac{az+b}{cz+d} \quad (a,b,c,d \in \mathbb{C}, ad - bc \neq 0),$$
from $\hat{\mathbb{C}}$ to $\hat{\mathbb{C}}$, has an extension to an isometry from $\mathcal{H}^3_+$ to $\mathcal{H}^3_+$. One way to see this is to break down α into a composition of maps of the form

(i) $z \to z + \lambda$ $(\lambda \in \mathbb{C})$, (ii) $z \to \mu z$ $(\mu \in \mathbb{C})$, (iii) $z \to -1/z$

and to extend these as follows on $\mathcal{H}^3_+$ (where z denotes $x_1 + ix_2$):

(i') $(z, x_3) \to (z + \lambda, x_3)$, (ii') $(z, x_3) \to (\mu z, |\mu| x_3)$,

$$\text{(iii')} \quad (z, x_3) \to \left(\frac{-\bar{z}}{|z|^2 + x_3^2}, \frac{x_3}{|z|^2 + x_3^2}\right).$$

To derive these expressions, observe that each of the maps (i), (ii) and (iii) above can be decomposed into two *inversions* (reflections) in circles in $\hat{\mathbb{C}}$. But each inversion has a unique extension to $\mathcal{H}^3_+$ as an inversion in the hemisphere bounded by the corresponding circle, and composing appropriate pairs of inversions gives us the formulae (i'), (ii') and (iii').

It is now an exercise (cf. Proposition 2.12) to show that $\mathrm{PSL}(2,\mathbb{C})$ preserves the metric ds on $\mathcal{H}^3_+$ and an exercise (cf. Proposition 2.13) to show that the geodesics are the arcs of semicircles as claimed. Moreover every orientation-preserving isometry of $\mathcal{H}^3_+$ is the extension of a conformal automorphism of $\hat{\mathbb{C}}$, since it must send hemispheres orthogonal to $\hat{\mathbb{C}}$ to hemispheres orthogonal to $\hat{\mathbb{C}}$, hence circles in $\hat{\mathbb{C}}$ to circles in $\hat{\mathbb{C}}$. Thus all orientation-preserving isometries of $\mathcal{H}^3_+$ are given by elements of $\mathrm{PSL}(2,\mathbb{C})$ acting as above.

The *Poincaré disc model* for hyperbolic three-space is the interior $\mathbb{D}^3$ of the unit disc in Euclidean three-space $\mathbb{R}^3$, equipped with the metric

$$ds = \frac{2\sqrt{(dx_1)^2 + (dx_2)^2 + (dx_3)^2}}{1 - (x_1^2 + x_2^2 + x_3^2)}.$$

Geodesics are arcs of circles orthogonal to the boundary sphere S^2.

5.1. *Types of isometries of hyperbolic 3-space*

A non-identity element $\alpha \in \mathrm{PSL}(2,\mathbb{C})$ has either two distinct fixed points on $\hat{\mathbb{C}}$ or a unique fixed point there, since $z = (az+b)/(cz+d)$ is a quadratic equation in z. If α has a unique fixed point then it is conjugate to $z \to z+1$ (just move the fixed point to ∞ and scale). If it has two distinct fixed points, it is conjugate to $z \to \lambda z$ for some $\lambda \in \mathbb{C}\setminus\{0,1\}$ (just move the fixed points to 0 and ∞), and we can distinguish between three types of behaviour corresponding to $|\lambda| = 1$, $\lambda \in \mathbb{R}\setminus\{-1,+1\}$, and $\lambda \in \mathbb{C}\setminus\mathbb{R}$.

We say that α is:

elliptic $\Leftrightarrow$ α fixes some geodesic in $\mathcal{H}^3_+$ pointwise;
parabolic $\Leftrightarrow$ α has a single fixed point in $\hat{\mathbb{C}}$;
hyperbolic $\Leftrightarrow$ α has two fixed points in $\hat{\mathbb{C}}$, and every hyperplane in $\mathcal{H}^3_+$ which contains the geodesic joining them is invariant under α;
loxodromic $\Leftrightarrow$ α has two fixed points in $\hat{\mathbb{C}}$, and no invariant hyperplane in $\mathcal{H}^3_+$.

5.2. The ordinary set of a Kleinian group

A *Kleinian group* is a *discrete* subgroup of $\mathrm{PSL}(2,\mathbb{C})$. Thus for a subgroup G of $\mathrm{PSL}(2,\mathbb{C})$ to be called Kleinian we require that there be no sequence $\{g_n\}$ of distinct elements of G tending to a limit $g \in \mathrm{PSL}(2,\mathbb{C})$ (here the topology on $\mathrm{PSL}(2,\mathbb{C})$ is that induced by the usual norm on the entries of a matrix). If G is discrete then for any $N > 0$ the number of elements of G having norm $\leq N$ is *finite*, since every infinite sequence with bounded norm has a convergent subsequence. So every discrete G is *countable*.

The action of G is *discontinuous* at $z \in \hat{\mathbb{C}}$ if there is a neighbourhood U of z in $\hat{\mathbb{C}}$ such that $g(U) \cap U = \emptyset$ for all but finitely many $g \in G$. (There is an analogous definition for the action of G on $\mathcal{H}_+^3$.)

Example. The modular group $G = \mathrm{PSL}(2,\mathbb{Z})$ acts discontinuously on $\hat{\mathbb{C}} \setminus \hat{\mathbb{R}}$, since when we tile $\hat{\mathbb{C}} \setminus \hat{\mathbb{R}}$ by copies of the standard fundamental domain, no more than six tiles meet at any one point.

The set of all $z \in \hat{\mathbb{C}}$ at which the action of G is discontinuous is called the *ordinary* (or *discontinuity*) set $\Omega(G)$. It follows at once from the definition that $\Omega(G)$ is *open* and *G-invariant*.

5.3. The action of a Kleinian group on $\mathcal{H}_+^3$

Theorem 5.1. *A subgroup G of $\mathrm{PSL}(2,\mathbb{C})$ is discrete if and only if it acts discontinuously on $\mathcal{H}_+^3$.*

Proof. If G is not discrete there exist $g_n \in G$ with limit $g \in \mathrm{PSL}(2,\mathbb{C})$. So for all $x \in \mathcal{H}_+^3$, $g_m^{-1} g_n(x) \to x$ as $m, n \to \infty$. Thus for any $x \in \mathcal{H}_+^3$ and neighbourhood U of x, for m and n sufficiently large $g_m^{-1} g_n(U) \cap U \neq \emptyset$. Hence G does not act discontinuously at x.

Conversely, if G does not act discontinuously at $x \in \mathcal{H}_+^3$, then for any neighbourhood U of x there exist a sequence $\{x_n\} \in U$ and (distinct) $g_n \in G$ such that each $g_n(x_n) \in U$. Take U compact. By passing to subsequences we may assume the x_n tend to a point y and the $g_n(x_n)$ tend to a point z (with both y and z in U). Now let k be an isometry of $\mathcal{H}_+^3$ having $k(z) = y$ and let $\{h_n\}, \{j_n\}$ be sequences of isometries, both tending to the identity, and having $h_n(y) = x_n$ and $j_n g_n(x_n) = z$ respectively. Consider $f_n = k j_n g_n h_n$. For each n this fixes y (by construction). But the isometries of $\mathcal{H}_+^3$ fixing a common point of $\mathcal{H}_+^3$ are a compact group. Hence the $\{f_n\}$ have a convergent subsequence and therefore so do the $\{g_n\}$. □

This theorem tells us that a Kleinian group G acts discontinuously on $\mathcal{H}_+^3$. One can construct a *fundamental polyhedron* for such an action, that is to say a (hyperbolic) polyhedron $\Delta_G \subset \mathcal{H}_+^3$ whose images under G *tile* $\mathcal{H}_+^3$: they intersect only in faces and their union is $\mathcal{H}_+^3$. The group G acts on Δ_G as face-pairing homeomorphisms and the quotient space $\Delta_G/G \cong \mathcal{H}_+^3/G$ has the structure of a *hyperbolic orbifold*: if G is a torsion-free group it is a *hyperbolic manifold*. Over the last thirty years there have been spectacular advances in the theory of these manifolds (see [16, 17]).

5.4. *Limit sets of Kleinian groups*

One can define the notion of the *limit set* $\Lambda(G)$ of a Kleinian group G in terms of its action either on $\mathcal{H}_+^3$, or on the boundary $\hat{\mathbb{C}}$ of $\mathcal{H}_+^3$. Definitions 5.2 and 5.3 below can be proved equivalent.

Definition 5.2. For $x \in \mathcal{H}_+^3$, let

$$\Lambda(x) = \{w \in \hat{\mathbb{C}} : w \text{ is an accumulation point of the orbit } Gx\}$$

where the topology on $\mathcal{H}_+^3 \cup \hat{\mathbb{C}}$ is that given by the Euclidean metric in the Poincaré disc model. Note that the Gx cannot have accumulation points in $\mathcal{H}_+^3$, since G acts discontinuously there. The set $\Lambda(x)$ is independent of $x \in \mathcal{H}_+^3$, as points a fixed hyperbolic distance apart get closer together in the Euclidean metric as one approaches the boundary of the Poincaré disc. Define $\Lambda(G)$ to be $\Lambda(x)$ for any $x \in \mathcal{H}_+^3$ or equivalently to be the set of all accumulation points (in the Poincaré disc model) of orbits of G on $\mathcal{H}_+^3$.

Notice that if G has a compact fundamental polyhedron $\Delta_G \subset \mathcal{H}_+^3$ then every point of $\hat{\mathbb{C}}$ is an accumulation point of tiny copies of Δ_G (in the Euclidean metric on the Poincaré disc), so $\Lambda(G)$ is the whole of $\hat{\mathbb{C}}$, whereas if Δ_G meets $\hat{\mathbb{C}}$ in a set with non-empty interior, the G-orbits of points in this set will be in $\hat{\mathbb{C}} \setminus \Lambda(G)$.

Definition 5.3. For $z \in \hat{\mathbb{C}}$, let

$$\Lambda(z) = \{w \in \hat{\mathbb{C}} : w \text{ is an accumulation point of the orbit } Gz\}$$

where the topology on $\hat{\mathbb{C}}$ is that of the spherical metric. It can be shown that $\Lambda(z)$ is independent of $z \in \hat{\mathbb{C}}$, except for a finite set of points z if G is *elementary* (see below). Define $\Lambda(G)$ to be $\Lambda(z)$ for generic $z \in \hat{\mathbb{C}}$, or equivalently to be the set of all accumulation points of orbits of G on $\hat{\mathbb{C}}$.

5.5. Elementary Kleinian groups

A Kleinian group G is called *elementary* if there exists a finite G orbit on either $\mathcal{H}^3_+$ or $\hat{\mathbb{C}}$. All elementary Kleinian groups G belong to the following three classes. For a proof see for example [3] or [22].

(i) G is conjugate to a finite subgroup of SO(3) acting on the Poincaré disc by rigid rotations fixing the origin (for example the symmetry group of a regular solid). In this case $\Lambda(G) = \emptyset$.
(ii) G is conjugate to a discrete group of Euclidean motions of $\mathbb{C}$ (i.e., fixing $\infty \in \hat{\mathbb{C}}$). (For example, the group generated by $z \to z+1$ and $z \to z+i$.) Then $|\Lambda(G)| = 1$.
(iii) G is conjugate to a group in which all elements are of the form $z \to kz$ or $z \to k/z$ for $k \in \mathbb{C}$. Then $|\Lambda(G)| = 2$.

The following proposition follows at once from this classification.

Proposition 5.4. *A Kleinian group G is elementary if and only if $|\Lambda(G)| \leq 2$, and non-elementary if and only if $\Lambda(G)$ is infinite.*

5.6. Properties of ordinary and limit sets

Theorem 5.5. *Every Kleinian group G acts discontinuously on $\hat{\mathbb{C}} \backslash \Lambda(G)$. Hence $\hat{\mathbb{C}}$ is the disjoint union of $\Omega(G)$ and $\Lambda(G)$.*

Proof. (Outline.) For groups G with $|\Lambda(G)| = 0, 1$ the result can be verified by checking the corresponding types of elementary Kleinian groups, so we may assume $|\Lambda(G)| \geq 2$. Now let $C(G)$ be the *convex hull* of $\Lambda(G)$ in $\mathcal{H}^3_+ \cup \hat{\mathbb{C}}$, the space obtained from $\mathcal{H}^3_+$ by 'scooping out' every open half 3-ball bounded by a round 2-disc contained in $\hat{\mathbb{C}} \backslash \Lambda(G)$. The set $C(G)$ is closed and G-invariant, since $\Lambda(G)$ is. There is a uniquely defined retraction:

$$\rho : \mathcal{H}^3_+ \cup (\hat{\mathbb{C}} \backslash \Lambda(G)) \to C(G)$$

sending each point of $\mathcal{H}^3$ to the nearest point of $C(G)$ (in the hyperbolic metric). This map ρ is continuous and commutes with the action of G. Now let z be any point of $\hat{\mathbb{C}} \backslash \Lambda(G)$ and $U \subset \hat{\mathbb{C}} \backslash \Lambda(G)$ be a neighbourhood of z. Then $\rho(U)$ is contained in a neighbourhood V of $\rho(z)$, and by taking U small (in the spherical metric) we can take V as small as we please (in the hyperbolic metric). But now since the action of G is discontinuous (by Theorem 5.1) V meets $g(V)$ for at most finitely many $g \in G$. Hence $g\rho(U)$

meets $\rho(U)$ for at most finitely many $g \in G$, and so $g(U)$ meets U for at most finitely many $g \in G$, in other words $z \in \Omega(G)$. □

Properties of $\Lambda(G)$

The following properties of $\Lambda(G)$ for non-elementary Kleinian groups G can be deduced from results stated so far. Note the similarities with properties of Julia sets of rational maps (Section 3.4 and Theorem 3.9):

(1) Every non-empty closed G-invariant subset S of $\hat{\mathbb{C}}$ contains $\Lambda(G)$;
(2) $\Lambda(G)$ is perfect (and hence, in particular, uncountable);
(3) Either $\Lambda(G) = \hat{\mathbb{C}}$ or $\Lambda(G)$ has empty interior.
(4) $\Lambda(G)$ is the closure of the set of fixed points of loxodromic and hyperbolic elements of G.

5.7. Analogies between rational maps and Kleinian groups: Sullivan's dictionary

Proposition 5.6. *Let G be a Kleinian group. Then $\Omega(G)$ is the largest open subset of $\hat{\mathbb{C}}$ on which the elements of G form an equicontinuous family.*

Proof. Assume G non-elementary (the elementary groups can be dealt with case by case). Then $\Lambda(G)$ contains at least three points (in fact infinitely many) so $\Omega(G)$ is contained in the equicontinuity set by Montel's theorem. But given any $z \in \Lambda(G)$, by property (4) of Section 5.6 there must be a repelling fixed point of some $g \in G$ arbitrarily close to z, so the family of maps G cannot be equicontinuous at z. □

We deduce the following (compare Proposition 3.7 and Corollary 3.8).

Proposition 5.7. *Let G be a non-elementary Kleinian group, and U be any open subset of $\hat{\mathbb{C}}$ meeting $\Lambda(G)$. Then*
$$\bigcup_{g \in G} gU = \hat{\mathbb{C}}.$$

Proof. The union $\bigcup_{g \in G} gU$ covers all of $\hat{\mathbb{C}}$ except at most two points (else the family G would be equicontinuous on U by Montel's theorem). So the complement of this union is a finite G-invariant set, and is therefore empty, as G is non-elementary. □

Corollary 5.8. *Let G be a non-elementary Kleinian group, and U be any open subset of $\hat{\mathbb{C}}$ meeting $\Lambda(G)$. Then*
$$\bigcup_{g \in G} g(U \cap \Lambda(G)) = \Lambda(G).$$

Sullivan's Dictionary is an enlightening (if not always precise) correspondence between definitions, conjectures and theorems in the realms of iterated rational maps and of Kleinian groups. Some entries are elementary, e.g., Julia set ↔ limit set, but some theorems on one side are conjectures on the other, and not everything works in exactly the same way. For example, we have the following.

Ahlfors' Conjecture, formulated by Ahlfors in the 1960s and proved by him for *geometrically finite* Kleinian groups, states in its most general form that for any finitely generated Kleinian group G either $\Lambda(G) = \hat{\mathbb{C}}$ or $\Lambda(G)$ has Lebesgue measure zero. This was finally proved in 2004 as a consequence of work by many authors (see [16, Theorem 5.6.6]).

Fatou's Question. Can the Julia set of a polynomial have positive Lebesgue measure? This question was finally answered in 2005 by Xavier Buff and Arnaud Chéritat, who proved that there exist quadratic polynomials, $z \to z^2 + c$, with positive area Julia sets. The proof is very technical, but see their paper at the 2010 ICM for an overview of the method.

As far as I know, there is no single location where all the current contents of this 'dictionary' are listed, but see Chapter 5 of [20] for the situation in 2000. More recently Richard Canary gave three talks on advanced topics in the dictionary at Dennis Sullivan's 70th birthday conference at Stony Brook in 2011: these can be found on the web.

6. The Family of Quadratic Polynomials $q_c : z \to z^2 + c$

Every quadratic polynomial is conformally conjugate to a unique polynomial of the form $q_c : z \to z^2 + c$ (Exercise 3, Section 9).

6.1. *The Mandelbrot set and the MLC conjecture*

The *Mandelbrot set* is defined as $M = \{c : J(q_c) \text{ connected}\} \subset \mathbb{C}$ (Fig. 6).

Theorem 6.1. *M is the set of c such that $q_c^n(0)$ does not tend to ∞.*

Proof. For any polynomial p of degree ≥ 2 the boundary ∂B_∞ of the basin of attraction of ∞ is the Julia set $J(p)$, since it is closed, completely invariant and disjoint from the Fatou set. But if $q_c^n(0)$ does not tend to ∞ then the only critical value in the basin of attraction B_∞ is ∞ itself, and so there is no obstruction to extending the Böttcher conjugacy (Theorem 4.4)

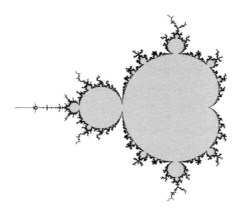

Fig. 6. The Mandelbrot set M.

from a neighbourhood of ∞ to the whole of B_∞. Hence B_∞ is homeomorphic to the open unit disc, and its complement $\hat{\mathbb{C}}\backslash B_\infty$ is therefore connected, as is their common boundary ∂B_∞. Conversely, if $q_c^n(0)$ tends to ∞ then $J(q_c)$ is totally disconnected by Proposition 3.10. □

Theorem 6.2 (Douady and Hubbard 1982). *M is connected.*

Proof. (Outline) The *filled Julia set* of q_c is $K(q_c) = \{z : q_c^n(z) \not\to \infty\}$. When $c \in M$, the Böttcher conjugacy (Theorem 4.4) defines a conformal bijection $\phi_c : \hat{\mathbb{C}}\backslash K(q_c) \to \hat{\mathbb{C}}\backslash \mathbb{D}$, conjugating q_c to $z \to z^2$. When $c \notin M$ the map ϕ_c, though not defined on the whole of the complement of K_c, is nevertheless defined on a neighbourhood of ∞ which contains the critical value c of q_c. For $c \in \hat{\mathbb{C}}\backslash M$ let $\Psi(c) = \phi_c(c)$. Then $\Psi : \hat{\mathbb{C}}\backslash M \to \hat{\mathbb{C}}\backslash \mathbb{D}$ is a conformal bijection (see [12]). Hence M is connected. □

A topological space X is called *locally connected* if every $x \in X$ has arbitrarily small connected (but not necessarily open) neighbourhoods (see [19, Section 17]).

Conjecture 6.3 ('MLC'). *M is locally connected.*

If MLC is true, then, by a theorem of Carathéodory, Ψ^{-1} extends to a continuous map from $\partial(\hat{\mathbb{C}}\backslash\mathbb{D}) = S^1$ onto the boundary ∂M. One consequence of MLC would be a purely combinatorial description of M.

A component of the interior of M is said to be *hyperbolic* if for every c in it the map q_c has an attracting or superattracting periodic orbit.

Conjecture 6.4 ('Hyperbolicity is dense'). *Every component of the interior of M is hyperbolic.*

Douady and Hubbard proved in [13] that MLC implies 'Hyperbolicity is dense'. There has been much progress since on both conjectures, but neither has been established in full generality. Yoccoz proved local connectivity at all but 'infinitely renormalisable points' of ∂M, and it has since been proved at many of these; 'hyperbolicity is dense' has been proved for components of M meeting the real axis (Lyubich, McMullen, Swiatek: see [18]).

6.2. The cardioid M_0

The most prominent component of the interior of M is the cardioid

$$M_0 = \{c : q_c \text{ has an attracting fixed point}\}.$$

$M_0 \cup \{1/4\}$ can be characterised as $\{c : J(q_c) \text{ is a topological circle}\}$.

Lemma 6.5. $M_0 = \{c : c = \lambda/2 - \lambda^2/4 \text{ for some } \lambda \text{ with } |\lambda| < 1\}$.

Proof. Consider the *logistic family* of quadratic maps: $p_\lambda(z) = \lambda z(1-z)$. It is easily seen that p_λ is conjugate to q_c if and only if $c = \lambda/2 - \lambda^2/4$. But the fixed points of p_λ have multipliers λ and $2 - \lambda$. The result follows (since $\lambda/2 - \lambda^2/4 = (2-\lambda)/2 - (2-\lambda)^2/4$). $\square$

Thus M_0 is parametrised by the multiplier λ of the attracting fixed point of q_c. The maps q_c with $c \in M_0 \setminus \{0\}$ are all topologically conjugate, indeed they are *quasiconformally* conjugate (see Section 7).

6.3. The intersection of M with the real axis

We consider how the behaviour of q_c varies for c on the real axis:

For $c > 1/4$, $J(q_c)$ is a Cantor set. At $c = 1/4$, q_c has a unique fixed point in $\mathbb{C}$. It is parabolic since it is at $z = 1/2$, so it has multiplier 1.

For $-3/4 < c < 1/4$, q_c has an attracting fixed point and $J(q_c)$ is a (topological) circle, with dynamics conjugate to that of the shift. At $c = -3/4$, the two points on the repelling period 2 orbit collide with the attracting fixed point, to form neutral fixed point (multiplier -1).

For $-5/4 < c < -3/4$, q_c has an attracting period 2 orbit, and the topology of $J(q_c)$ is the same as that for the (superattractive) case $c = -1$.

For $-2 < c < -5/4$, as c decreases through this range, starting with a sequence of period doublings we have the whole Milnor–Thurston sequence of periods for real unimodal maps, familiar to dynamicists [11]. At $c = -2$, $J(q_c)$ is the real interval $[-2, +2]$ (Exercise 2, Section 9).

For $c < -2$, the Julia set is again a Cantor set.

6.4. *Internal and external rays*

The *internal ray* of argument ν in the component M_0 is the set of $c \in M_0$ for which the multiplier λ of q_c at its fixed point has argument $2\pi\nu$. When c crosses ∂M_0 at the point where $\lambda = e^{2\pi i p/q}$, it passes into an adjoining component of int(M) where the fixed point becomes an attracting cycle of period q. There are further bifurcations as we continue along any path through different components of int(M).

We use the combinatorics of 'external rays' to give us information about the structure of M. When $c \in M$, for any $\theta \in [0,1)$, the image of the radial line $\arg(z) = 2\pi\theta$ on $\hat{\mathbb{C}}\backslash\mathbb{D}$ under the inverse ϕ_c^{-1} of the Böttcher conjugacy is called the *external ray* $\mathcal{R}_\theta$ of argument $2\pi\theta$ on $\hat{\mathbb{C}}\backslash K(q_c)$. Similarly, in the parameter plane, the *external ray* $\mathcal{R}_\theta$ of argument $2\pi\theta$ on $\hat{\mathbb{C}}\backslash M$ is the image of the radial line $\arg(z) = 2\pi\theta$ on $\hat{\mathbb{C}}\backslash\mathbb{D}$ under the inverse Ψ^{-1} of the Douady–Hubbard map. A ray is said to *land* if it accumulates at a unique point of $J(q_c)$ (in the dynamical case) or ∂M (in the parameter case).

Theorem 6.6 (Douady and Hubbard). *Every parameter space external ray with rational angle θ lands at a point c of ∂M. If θ has odd denominator then q_c has a parabolic cycle. If θ has even denominator then the critical point 0 of q_c is strictly preperiodic.*

Proof. See [13] or [19] or [10]. □

If $J(q_c)$ (or ∂M respectively) is locally connected then all external rays land, by Carathéodory's criterion. For certain $c \in \partial M$ there are non-landing dynamical external rays (so these $J(q_c)$ are not locally connected). The conjecture MLC is equivalent to all parameter external rays landing.

Which rational external rays land on ∂M_0? The endpoint on ∂M_0 of the internal ray of argument $\nu = 1/3$ is $c = (-1 + 3\sqrt{3}i)/8$. The external rays $1/7, 2/7, 4/7$ in the dynamical plane landing at the fixed point α of this q_c are shown in Fig. 7.

The arguments of the two rays which together enclose the component of the interior of $K(q_c)$ containing the critical value are $\theta_-(1/3) = 1/7 = .\overline{001}$ (in binary) and $\theta_+(1/3) = 2/7 = .\overline{010}$. In parameter space the external rays

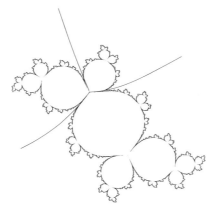

Fig. 7. External rays landing at the fixed point α of Douady's 'fat rabbit'.

with the same arguments, $\theta_-(1/3)$ and $\theta_+(1/3)$, land at $c \in \partial M_0$. We have a similar picture for any rational p/q in place of $1/3$. Here there is a q-cycle of external rays landing at the fixed point. The angles $\theta_\pm(p/q)$ of the pair enclosing the critical value are given by a combinatorial algorithm (see [9]) due originally to Hedlund and Morse in their pioneering work on symbolic dynamics in the 1930s.

The combinatorial structure of Julia sets $J(q_c)$ for $c \in M$ is modelled using *Hubbard trees* (Douady and Hubbard [13]) or *quadratic invariant laminations* on the unit disc (see Part II of Thurston's paper in [23], and its appendix by Schleicher: this paper circulated widely as a preprint for more than two decades). Shrinking the leaves of an invariant lamination to points yields a 'pinched disc' model for the corresponding filled Julia set. The counterpart in parameter space is Lavaurs' algorithm [15], which constructs a *combinatorial Mandelbrot set* as the quotient of the unit disc by an equivalence relation. If MLC holds, M is homeomorphic to this set.

7. Quasiconformal Mappings: The Measurable Riemann Mapping Theorem and its Applications

We conclude with a very brief introduction to a key technical tool used in many constructions and proofs for both rational maps and Kleinian groups.

Given any two Riemann surfaces $\mathcal{S}_1, \mathcal{S}_2$ which are homeomorphic to a sphere, there is a *conformal* homeomorphism $\mathcal{S}_1 \to \mathcal{S}_2$. This is not true for

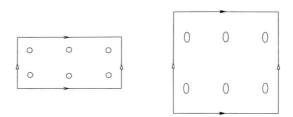

Fig. 8. There is a quasiconformal homeomorphism between these two tori.

Riemann surfaces of genus > 0, for example tori, but it becomes true if we weaken the requirement of conformality to 'quasiconformality' (Fig. 8).

A homeomorphism f between open sets in $\mathbb{C}$ is said to be *K-quasiconformal* if it sends infinitesimally small round circles to infinitesimally small ellipses which have ratio of semi-major axis length to semi-minor axis length less than or equal to K. Writing f as a function of z and $\bar{z}$ (if f were conformal it would be function of z only), we can associate to f the *Beltrami form*:

$$\mu(f) = \frac{\partial f}{\partial \bar{z}} \Big/ \frac{\partial f}{\partial z}$$

and it is straightforward to prove that f is K-quasiconformal, with $K = (1+k)/(1-k)$, if and only if $\mu(f)$ is defined almost everywhere and has essential supremum $\|\mu\|_\infty \le k < 1$.

The *Riemann Mapping Theorem* asserts that if U is a bounded open simply-connected subset of $\mathbb{C}$, then there exists a conformal orientation-preserving homeomorphism $\phi : U \to \mathbb{D}$, unique up to post-composition by orientation-preserving conformal homeomorphisms of $\mathbb{D}$. The *Measurable Riemann Mapping Theorem*, due to Morrey, Bojarski, Ahlfors and Bers, asserts an analogous result when we are also given a measurable function $U \to \mathbb{C}$ with $\|\mu\|_\infty = k < 1$. It states there exists a *quasiconformal* homeomorphism $f : U \to \mathbb{D}$ which has Beltrami derivative the given $\mu(z)$ at (almost) every $z \in U$. There are various forms of this theorem, which is also referred to as the 'Integrability Theorem': see [6] or Douady's paper in [24] for precise statements. We list some of the many applications:

(1) *Quasiconformal conjugacy of quadratic polynomials.* Any two maps in the same hyperbolic component of the interior of M, but with neither corresponding to the *centre* of the component, are qc conjugate.

(2) *Bers' Simultaneous Uniformisation Theorem* [5]. Suppose G_1 and G_2 are geometrically finite topologically conjugate Fuchsian groups (discrete subgroups of $\mathrm{PSL}_2(\mathbb{R})$), both with limit set $\hat{\mathbb{R}} \subset \hat{\mathbb{C}}$. Bers' Simultaneous Uniformisation Theorem tells us that there exists a discrete subgroup G of $\mathrm{PSL}_2(\mathbb{C})$, such that: the limit set is a quasicircle $\Lambda \subset \hat{\mathbb{C}}$; on one component of $\hat{\mathbb{C}} \setminus \Lambda$ the action of G is conformally conjugate to the action of G_1 on the upper half-plane; on the other component the action of G is conformally conjugate to that of G_2 on the lower half-plane.

(3) *Polynomial-like mappings.* In [14], Douady and Hubbard introduced the notion of a *polynomial-like map*, a proper holomorphic surjection $p : V \to U$ where U and V are simply-connected open sets in $\mathbb{C}$ with $U \supset \overline{V}$. This has a well-defined *filled Julia set* $K(p) = \bigcap_{n \geq 0} p^{-n}(\overline{V})$. The *Straightening Theorem* of Douady and Hubbard states that for every polynomial-like mapping p there is a genuine polynomial map P which is hybrid equivalent to p (where *hybrid equivalent* means that there is a quasiconformal conjugacy h between p and P on neighbourhoods of $K(p)$ and $K(P)$ such that the Beltrami form of h vanishes on $K(p)$).

(4) *Tuning, renormalisation and baby Mandelbrot sets.* Often q_c has the property that there is some region of the dynamical plane for which the first return map $(q_c)^n$ is a quadratic-like map, and is therefore hybrid equivalent to some $q_{c'}$ by the Straightening Theorem. The Julia set of q_c then contains copies of $J(q_{c'})$. Families of quadratic-like first return maps satisfying certain conditions give rise to 'baby Mandelbrot sets' — small copies of M. Indeed, every point of ∂M is an accumulation point of baby Mandelbrot sets (see McMullen's article in [24]).

(5) *Matings of polynomials.* Quasiconformal surgery methods can be used to *mate* q_c with $q_{c'}$ for any pair c, c' in the interior of the cardioid M_0, that is to construct a *rational* map q of degree two which has Julia set $J(q)$ a quasicircle and which is conformally conjugate to q_c on one component of $\hat{\mathbb{C}} \setminus J(q)$ and to $q_{c'}$ on the other. Mary Rees and Tan Lei proved that any two *hyperbolic* quadratic polynomials $q_c, q_{c'}$ can be mated if and only if c' is not in the conjugate limb of M to that of c.

(6) *Matings of rational maps with Kleinian groups.* If we extend our notions of rational maps and Kleinian groups to include *holomorphic correspondences*, it becomes possible to mate a hyperbolic quadratic polynomial with the modular group [6, 8].

8. Further Reading

Beardon [4] and Milnor [19] are very readable accounts of the theory of iterated rational maps (the debt owed to them by the author of this chapter will be apparent). Branner and Fagella [6] on quasiconformal surgery is both enlightening and comprehensive. As far as Kleinian groups and hyperbolic geometry are concerned, we have hardly begun here. For the basic theory see Beardon [3] and Ratcliffe [22]. The quotient of hyperbolic 3-space by a torsion-free Kleinian group is a hyperbolic 3-manifold: see [16, 17] for a comprehensive account of the remarkable results proved about such manifolds in recent years. Finally, the inspirational books of Thurston [25] and of Mumford, Series and Wright [21] are a pleasure to read.

9. Exercises

(1) For the angle-doubling map $t \to 2t$ mod 1 on the circle $\mathbb{R}/\mathbb{Z}$ prove that the periodic points are the points $t \in [0, 1)$ of the form $t = m/(2^n - 1)$ (where $0 \le m < 2^n - 1$ with $m, n \in \mathbb{N}$).

(2) Show that $h : z \to z + 1/z$ is a semiconjugacy from $f : z \to z^2$ to $g : z \to z^2 - 2$ (that is, h is a surjection satisfying $hf = gh$) and that h sends the Julia set of f (the unit circle) onto the real interval $[-2, +2]$.

(3) Prove that every quadratic map $f(z) = \alpha z^2 + \beta z + \gamma$ with $\alpha \ne 0$ is conjugate to $q_c(z) = z^2 + c$ for a unique c. *Hint: the conjugating map must send ∞ to ∞, and hence be of the form $h(z) = kz + l$.*

(4) Find all critical points, and their forward orbits, for the rational map

$$f : z \to \frac{-2z - 1}{z^2 + 4z + 2}.$$

Deduce that f is conjugate to $z \to z^2 - 1$.

(5) If f is a rational function with a fixed point at ∞ show that the multiplier of f at ∞ is equal to $\lim_{z \to \infty} 1/f'(z)$. Deduce that the fixed point at ∞ is a critical point if and only if $\lim_{z \to \infty} f'(z) = \infty$.

(6) On the hyperbolic plane a *reflection* is an orientation-reversing isometry β which fixes some geodesic pointwise.

   (i) Show that every reflection β is an *involution* (i.e., $\beta^2 = I$);
   (ii) Show that for every reflection β there is an element of $\mathrm{PSL}(2, \mathbb{R})$ which conjugates β to 'reflection in the imaginary axis', $z \to -\bar{z}$.

(iii) Show that every orientation-preserving isometry of the hyperbolic plane is the composition of a pair of reflections (by Section 5.1 it will suffice to consider $z \to \lambda z$ and $z \to z \pm 1$ on $\mathcal{H}_+$, and $z \to e^{i\theta} z$ on $\mathbb{D}$).

(7) Prove that a Möbius transformation of the form
$$A = \begin{pmatrix} a & b \\ b & d \end{pmatrix} \text{ where } ad - b^2 = 1$$
satisfies $JAJ = A^{-1}$ where $J(z) = -z^{-1}$. Deduce that if a discrete group G is generated by elements of this form then $\Lambda(G)$ is invariant under J.

(8) Let C_1, C_2 and C_3 be circles in the plane which touch in pairs and have disjoint interiors, and let R_j denote reflection in C_j. It can be proved that the group of orientation-preserving conformal automorphisms generated by $R_2 R_1$ and $R_3 R_2$ is discrete and has limit set the circle through the contact points of the pairs. Add a fourth circle C_4, which touches C_1, C_2 and C_3 and has interior disjoint from theirs. Show that the limit set of the subgroup of $\mathrm{PSL}(2,\mathbb{C})$ generated by the $R_j R_k$ ($j, k \in \{1, 2, 3, 4\}$) is an Apollonian circle-packing (Fig. 3(c)), i.e., a circle-packing obtained from three initial pairwise touching circles by iteratively adding a new circle of maximal radius in each space between three pairwise touching circles already drawn.

(9) Show that for any $a \in \mathbb{D}$ the map:
$$\phi_a(z) = \frac{z - a}{1 - \bar{a} z}$$
carries the unit circle to itself, and the origin to a point of $\mathbb{D}$, and hence carries the unit disc $\mathbb{D}$ isomorphically to itself. A finite product of form
$$f(z) = e^{i\theta} \phi_{a_1}(z) \phi_{a_2}(z) \ldots \phi_{a_n}(z)$$
with $a_1, \ldots, a_n \in \mathbb{D}$, is called a *Blaschke product*. Show that f is a rational map which carries $\mathbb{D}$ onto $\mathbb{D}$ and $\hat{\mathbb{C}} \setminus \bar{\mathbb{D}}$ onto $\hat{\mathbb{C}} \setminus \bar{\mathbb{D}}$. Deduce that the unit circle S^1 is completely invariant and hence that $J(f) \subseteq S^1$.

(10) (*Shimizu's Lemma*) Prove that if G is a discrete subgroup of $\mathrm{PSL}(2,\mathbb{R})$ and
$$A = \begin{pmatrix} 1 & 1 \\ 0 & 1 \end{pmatrix} \in G \text{ and } B = \begin{pmatrix} a & b \\ c & d \end{pmatrix} \in G$$
then: (i) either $c = 0$ or $|c| \geq 1$; (ii) $|\mathrm{trace}(ABA^{-1}B^{-1}) - 2| \geq 1$.

Hint: Let $B_0 = B$ and $B_{n+1} = B_n A B_n^{-1}$. Show that if $|c| < 1$ then $B_n \to A$ as $n \to \infty$, contradicting discreteness. (See [7], which has an early plot of the Mandelbrot set in this context.)

10. Outline Solutions to Exercises (3), (5), (7) and (9)

(3) The map $z \to kz + l$ is the desired conjugacy if and only if we have $k(\alpha z^2 + \beta z + \gamma) + l = (kz + l)^2 + c$ for all $z \in \mathbb{C}$. That is, if and only if $k = \alpha$, $l = \beta/2$, and $c = k\gamma + l - l^2 = \alpha\gamma + \beta/2 - (\beta/2)^2$.

(5) Let $g = \sigma f \sigma$ where $\sigma(z) = 1/z$. Then the multiplier of f at its fixed point ∞ is (by definition) the multiplier of g at its fixed point 0. The Taylor series for g at $z = 0$ has the form $g(z) = \sum_{n=1}^{\infty} a_n z^n$ with a_1 the multiplier of g at $z = 0$. As $f(z) = \sigma g \sigma(z)$, we have, for large z,

$$f(z) = (a_k z^{-k} + a_{k+1} z^{-(k+1)} + a_{k+2} z^{-(k+2)} + \cdots)^{-1} = a_k^{-1} z^k (1 + \cdots),$$

where a_k is the first non-zero a_n and '$(1 + \cdots)$' is a power series in z^{-1}. Hence $\lim_{z \to \infty} f'(z) = a_1^{-1}$ if $a_1 \neq 0$, and $= \infty$ if $a_1 = 0$.

(7)
$$J(AJ(z)) = J\left(\frac{-a/z + b}{-b/z + d}\right) = \frac{b/z - d}{-a/z + b} = \frac{dz - b}{-bz + a} = A^{-1}(z).$$

If $z \in \Lambda(G)$ then for any point $z' \in \hat{\mathbb{C}}$ there is a sequence of elements $g_n \in G$ with $g_n(z')$ converging to z. But then, noting that $J^{-1} = J$, we deduce that the points $(Jg_n J)(J(z'))$ converge to $J(z)$. But $J(g_n)J = g_n^{-1} \in G$, so this means that $J(z)$ is an accumulation point of a G-orbit. Hence $J(z) \in \Lambda(G)$.

(9) We know that $z \in S^1$ if and only if $z = e^{i\theta}$ for some $\theta \in \mathbb{R}$. But

$$\phi_a(e^{i\theta}) = \frac{e^{i\theta} - a}{1 - \bar{a}e^{i\theta}} = \frac{e^{i\theta/2} - ae^{-i\theta/2}}{e^{-i\theta/2} - \bar{a}e^{i\theta/2}} \in S^1.$$

The map ϕ_a is a bijection, and since it sends S^1 to S^1, and 0 to $-a \in \mathbb{D}$ it must send the component $\mathbb{D}$ of $\hat{\mathbb{C}} \setminus S^1$ to itself. Since a product of complex numbers having modulus < 1 also has modulus < 1, we deduce that $f(\mathbb{D}) \subseteq \mathbb{D}$. Similarly $f(\hat{\mathbb{C}} \setminus \bar{\mathbb{D}}) \subseteq \hat{\mathbb{C}} \setminus \bar{\mathbb{D}}$ and so S^1 is completely invariant. As the Julia set $J(f)$ is the minimal closed completely invariant set, we deduce that $J(f) \subseteq S^1$. Or apply Montel's Theorem to the iterates of f on $\mathbb{D} \cup (\hat{\mathbb{C}} \setminus \bar{\mathbb{D}})$, to show that $\mathbb{D} \cup (\hat{\mathbb{C}} \setminus \bar{\mathbb{D}}) \subseteq F(f)$ and so $J(f) \subseteq S^1$.

References

[1] L. Ahlfors, *Complex Analysis*. McGraw-Hill (1966).
[2] M. Audin, *Fatou, Julia, Montel, le grand prix des sciences mathématiques de 1918, et après*. Springer (2009).
[3] A. Beardon, *The Geometry of Discrete Groups*. Graduate Texts in Mathematics, No. 91. Springer (1983).
[4] A. Beardon, *Iteration of Rational Functions*. Graduate Texts in Mathematics, No. 132. Springer (1991).
[5] L. Bers, Simultaneous uniformization. *Bull. Amer. Math. Soc.* **66**, 94–97 (1960).
[6] B. Branner and N. Fagella, *Quasiconformal Surgery in Holomorphic Dynamics*. Cambridge University Press (2014).
[7] R. Brooks and P. Matelski, The dynamics of 2-generator subgroups of PSL(2, $\mathbb{C}$). In *Riemann Surfaces and Related Topics: Proceedings of the 1978 Stony Brook Conference*. Princeton University Press (1980).
[8] S. Bullett and C. Penrose, Mating quadratic maps with the modular group. *Invent. Math.* **115**, 483–511 (1994).
[9] S. Bullett and P. Sentenac, Ordered orbits of the shift, square roots, and the devil's staircase. *Math. Proc. Cambridge Philos. Soc.* **115**, 451–481 (1994).
[10] L. Carleson and T. Gamelin, *Complex Dynamics*. Springer (1993).
[11] W. de Melo and S. van Strien, *One-Dimensional Dynamics*. Springer (1993).
[12] A. Douady and J. Hubbard, Itération des polynômes quadratiques complexes. *C. R. Acad. Sci. Paris* **294**, 123–126 (1982).
[13] A. Douady and J. Hubbard, *Étude dynamique des polynômes complexes I/II*. Publications Mathématiques d'Orsay (1984/85).
[14] A. Douady and J. Hubbard, On the dynamics of polynomial-like mappings. *Ann. Sci. École Norm. Sup.* **18**, 287–343 (1985).
[15] P. Lavaurs, Une description combinatoire de l'involution define par M sur les rationnels à denominateur impair. *C. R. Acad. Sci. Paris* **303**, 143–146 (1986).
[16] A. Marden, *Outer Circles: An Introduction to Hyperbolic 3-Manifolds*. Cambridge University Press (2007).
[17] A. Marden, *Hyperbolic Manifolds*. Cambridge University Press (2016).
[18] C. T. McMullen, *Complex Dynamics and Renormalization*. Annals of Math. Studies, No. 135. Princeton University Press (1994).
[19] J. Milnor, *Dynamics in One Complex Variable*. Annals of Mathematics Studies, No. 160. Princeton University Press (2006).
[20] S. Morosawa, Y. Nishimura, M. Taniguchi and T. Ueda, *Holomorphic Dynamics*. Cambridge University Press (2000).
[21] D. Mumford, C. Series and D. Wright, *Indra's Pearls, the Vision of Felix Klein*. Cambridge University Press (2002).
[22] J. G. Ratcliffe, *Foundations of Hyperbolic Manifolds*. Graduate Texts in Mathematics, Vol. 149. Springer (1994).

[23] D. Schleicher (ed.), *Complex Dynamics: Families and Friends*. A. K. Peters (2009).
[24] L. Tan (ed.), *The Mandelbrot Set, Theme and Variations*. LMS Lecture Notes, No. 274. Cambridge University Press (2000).
[25] W. Thurston, *Three-Dimensional Geometry and Topology*. Princeton University Press (1997).
[26] J.-C. Yoccoz, Petits Diviseurs en Dimension 1. *Astérisque* **231** (1995).

Chapter 5

Minimal Surfaces and the Bernstein Theorem

T. Bourni

*Fachbereich Mathematik und Informatik, Institut für Mathematik
Freie Universität Berlin
Arnimallee 3, 14195 Berlin, Germany
bourni@zedat.fu-berlin.de*

G. Tinaglia

*Department of Mathematics, King's College London
The Strand, London WC2R 2LS, UK
giuseppe.tinaglia@kcl.ac.uk*

The content of this chapter revolves around a very classical theorem in the theory of minimal surfaces in Euclidean space, namely the Bernstein theorem, and its proof. With the proof of the Bernstein theorem in mind, we then explore several properties of minimal surfaces more deeply than is required by the proof. Furthermore, we look into generalizations of the Bernstein theorem.

1. Introduction

This chapter is based on a series of lectures given by the first author at the Freie Universität Berlin as part of a module for the Berlin Mathematical School, and by the second author at King's College London as part of a module for the London Taught Course Centre.

The theory of minimal surfaces is a very classical theory that overlaps with many fields of mathematics. While several of the results presented in this chapter generalize to higher dimension, higher codimension and different ambient manifolds, in order to keep the notes self-contained and accessible to a wide audience of graduate students, we deal with minimal surfaces in Euclidean space and only refer to more general results.

There are many books written on this subject. For example, an interested reader could consult [5, 9, 12]. Indeed this chapter contains several of the arguments presented in [5].

1.1. *Minimal surfaces*

Let M be a smooth orientable surface (possibly with boundary) and let $x : M \to \mathbb{R}^3$ be an isometric immersion. In other words, the metric on M is the one induced by $\mathbb{R}^3$ (unless otherwise stated, we will assume that M is connected). We denote by N the unit normal vector field to M,

$$N \colon M \to \mathbb{S}^2,$$

and by T_pM the tangent plane to M at p. Given $p \in M$, let

$$d_p x \colon T_pM \to T_{x(p)}x(M)$$

denote the differential of x at p. Abusing the notation, we will often not distinguish between T_pM and $T_{x(p)}x(M)$ and we will use the following notation

$$d_p x X = x_*(p)X = Xx(p).$$

In fact, we will usually simplify the notation above by omitting p. The shape operator at p is the differential

$$d_p N \colon T_pM \to T_{N(p)}\mathbb{S}^2 \cong T_pM.$$

The second fundamental form at p is the map

$$A_p \colon T_pM \to T_pM$$

given by

$$d_p N = -d_p x \circ A_p.$$

The principal curvatures at $p \in M$ are the eigenvalues of A_p which we denote by k_1, k_2. The mean curvature, H, is half of the trace of A, i.e.,

$$H = \tfrac{1}{2}(k_1 + k_2).$$

The Gaussian curvature, K, is given by the determinant of A, so

$$K = k_1 k_2.$$

The norm of A, $|A|$, is given by the square root of the trace of A^2, hence

$$|A| = \sqrt{k_1^2 + k_2^2}.$$

Note that
$$4H^2 = |A|^2 + 2K.$$

Definition 1.1 (First definition). An isometric immersion $x : M \to \mathbb{R}^3$ is a *minimal* isometric immersion if $H \equiv 0$. Then, $x(M)$ is a minimal surface.

Abusing the language, it will often happen that we will refer to M as a minimal surface.

We now choose coordinates (u, v) for M and do explicit computations. The tangent vectors $\frac{\partial}{\partial u} = \partial_u$ and $\frac{\partial}{\partial v} = \partial_v$ form a basis of the tangent space. In the following computation, given two vectors $v, w \in \mathbb{R}^3$, $\langle v, w \rangle$ denotes their inner product. Note that throughout these notes, we will also use $v \cdot w$ to denote such inner product. The unit normal is

$$N = \frac{\partial_u x \times \partial_v x}{|\partial_u x \times \partial_v x|}.$$

In coordinates (u, v) the metric is given by the matrix

$$\mathcal{F}_I = \begin{pmatrix} e & f \\ f & g \end{pmatrix},$$

where

$$e = \langle \partial_u, \partial_u \rangle = \langle \partial_u x, \partial_u x \rangle, \quad f = \langle \partial_u, \partial_v \rangle = \langle \partial_u x, \partial_v x \rangle,$$

and

$$g = \langle \partial_v, \partial_v \rangle = \langle \partial_v x, \partial_v x \rangle.$$

The second fundamental form is given by the matrix

$$\mathcal{F}_{II} = \begin{pmatrix} l & m \\ m & n \end{pmatrix},$$

where

$$l = \langle A\partial_u, \partial_u \rangle = -\langle \partial_u N, \partial_u x \rangle = \langle N, \partial_u \partial_u x \rangle,$$
$$m = \langle A\partial_u, \partial_v \rangle = \langle A\partial_v, \partial_u \rangle = -\langle \partial_u N, \partial_v x \rangle = -\langle \partial_v N, \partial_u x \rangle = \langle N, \partial_u \partial_v x \rangle$$

and

$$n = \langle A\partial_v, \partial_v \rangle = -\langle \partial_v N, \partial_v x \rangle = \langle N, \partial_v \partial_v x \rangle.$$

The mean curvature is half of the trace of $\mathcal{F}_I^{-1} \mathcal{F}_{II}$ while the Gaussian curvature is its determinant.

Examples

- The plane has a constant normal vector, hence the second fundamental form vanishes and consequently the mean curvature is zero.
- The helicoid is given in parametric form by

$$x(\theta, t) = (t\cos\theta, t\sin\theta, \theta).$$

A basis for the tangent space is given by

$$\partial_t = (\cos\theta, \sin\theta, 0),$$

$$\partial_\theta = (-t\sin\theta, t\cos\theta, 1).$$

Consequently,

$$N = \frac{\partial_t \times \partial_\theta}{|\partial_t \times \partial_\theta|} = \frac{(\sin\theta, -\cos\theta, t)}{\sqrt{1+t^2}}.$$

$$\partial_\theta N = \frac{(\cos\theta, \sin\theta, 0)}{\sqrt{1+t^2}} = \frac{1}{\sqrt{1+t^2}} \partial_t$$

$$\implies A\partial_\theta = -\frac{1}{\sqrt{1+t^2}} \partial_t,$$

$$\partial_t N = \frac{(0,0,1)}{\sqrt{1+t^2}} - \frac{t}{(1+t^2)^{\frac{3}{2}}} (\sin\theta, -\cos\theta, t)$$

$$= \frac{(-t\sin\theta, t\cos\theta, 1)}{(1+t^2)^{\frac{3}{2}}}$$

$$= \frac{\partial_\theta}{(1+t^2)^{\frac{3}{2}}}$$

$$\implies A\partial_t = -\frac{1}{(1+t^2)^{\frac{3}{2}}} \partial_\theta.$$

Thus, it is now easy to see that $H = 0$.

1.2. *First variation of area*

In this section we will obtain an equivalent definition of minimal immersions related to the area of a surface.

Definition 1.2. Given an isometric immersion $x : M \to \mathbb{R}^3$, a *variation of* $x : M \to \mathbb{R}^3$ with compact support and fixed boundary (if $\partial M \neq \emptyset$) is a smooth one-parameter family of isometric immersions

$$x_t \colon M \to \mathbb{R}^3, \quad t \in (-\epsilon, \epsilon)$$

such that $x_0 = x$ and for any $t \in (-\epsilon, \epsilon)$ we have that $x_t = x_0$ outside some compact set $W \subset M$ and that $x_t|_{\partial M} = x_0|_{\partial M}$ (if $\partial M \neq \emptyset$).

Unless otherwise stated, a variation of $x \colon M \to \mathbb{R}^3$ will always be a variation with compact support and fixed boundary.

Theorem 1.3 (First variation formula). *Let $x_t \colon M \to \mathbb{R}^3$ be a variation of $x \colon M \to \mathbb{R}^3$ and let $M_t := x_t(M)$. Then the first variation of the area of M along the variation x_t is given by*

$$\left.\frac{d}{dt} \operatorname{Area}(M_t)\right|_{t=0} = -2\int_M H\langle \dot{x}, N\rangle, \tag{1.1}$$

where $\dot{x} := \frac{dx_t}{dt}\big|_{t=0}$.

Before proving the first variation formula, we recall two definitions.

Definition 1.4. Given an isometric immersion $x \colon M \to \mathbb{R}^3$, $x \colon M \to \mathbb{R}^3$ is a critical point for the area functional if the first variation of the area of M along any variation is zero.

Definition 1.5. Let $V \colon M \to \mathbb{R}^3$ be a vector field, let $p \in M$ and let e_i, $i = 1, 2$, be an orthonormal frame defined on a neighbourhood of p in M. Then the *divergence over M* of V around p is

$$\operatorname{div}_M V = \sum_i \langle e_i(V), e_i\rangle.$$

If we write $V = V^{tg} + \langle V, N\rangle N$, where V^{tg} is the component of V tangent to M, then we obtain that

$$\begin{aligned}
\operatorname{div}_M V &= \sum_i \langle e_i V, e_i\rangle \\
&= \sum_i \langle e_i V^{tg} + e_i(\langle V, N\rangle N), e_i\rangle \\
&= \sum_i \langle \nabla_{e_i} V^{tg} + \langle Ae_i, V^{tg}\rangle N + e_i(\langle V, N\rangle) N - \langle V, N\rangle Ae_i, e_i\rangle \\
&= \sum_i \langle \nabla_{e_i} V^{tg} - \langle V, N\rangle Ae_i, e_i\rangle \\
&= \operatorname{div}_M V^{tg} - 2H\langle V, N\rangle.
\end{aligned}$$

Proof of Theorem 1.3. Let $g_{ij}(t)$ be the metric on M induced by the immersion x_t; then

$$\text{Area}(M_t) = \int_M \frac{\sqrt{\det(g_{ij}(t))}}{\sqrt{\det(g_{ij}(0))}}.$$

Thus,

$$\frac{d}{dt}\text{Area}(M_t) = \frac{d}{dt}\int_M \frac{\sqrt{\det(g_{ij}(t))}}{\sqrt{\det(g_{ij}(0))}} = \int_M \frac{d}{dt}\frac{\sqrt{\det(g_{ij}(t))}}{\sqrt{\det(g_{ij}(0))}}.$$

In order to compute $\frac{d}{dt}\sqrt{\det(g_{ij}(t))}|_{t=0}$ let us write the first-order Taylor expansion of x_t, namely

$$x_t = x + t\dot{x} + O(t^2).$$

Since we are only considering the first derivative at $t = 0$, in the following computation we will ignore any term where a t^2-term appears. Let e_i be an orthonormal frame for M. A tangent frame for M_t is

$$f_i = e_i x_t = e_i + t e_i(\dot{x}).$$

and the metric induced by x_t is

$$\begin{aligned} g_{ij}(t) &= \langle f_i, f_j \rangle \\ &= \langle e_i + t e_i(\dot{x}), e_j + t e_j(\dot{x}) \rangle \\ &= \delta_{ij} + t(\langle e_i(\dot{x}), e_j \rangle + \langle e_j(\dot{x}), e_i \rangle). \end{aligned}$$

Letting C denote the matrix with coefficients $C_{ij} := \langle e_i(\dot{x}), e_j \rangle + \langle e_j(\dot{x}), e_i \rangle$ and using the formula $\det(I + tC) = I + t\,\text{trace}(C) + O(t^2)$ we obtain

$$\begin{aligned} \det(g_{ij}(t)) &= \det(\delta_{ij} + tC) \\ &= 1 + t\,\text{trace}(C) \\ &= 1 + 2t\langle e_i(\dot{x}), e_i \rangle \\ &= 1 + 2t\,\text{div}_M(\dot{x}) \\ &= 1 + 2t\,\text{div}_M(\dot{x}^{tg}) - 4tH\langle \dot{x}, N \rangle. \end{aligned}$$

Finally, using Taylor expansion, we obtain

$$\begin{aligned} \sqrt{\det(g_{ij}(t))} &= \sqrt{1 + 2t\,\text{div}_M(\dot{x}^{tg}) - 4tH\langle \dot{x}, N \rangle} \\ &= 1 + t\,\text{div}_M(\dot{x}^{tg}) - 2tH\langle \dot{x}, N \rangle \end{aligned}$$

and
$$\left.\frac{d}{dt}\sqrt{\det(g_{ij}(t))}\right|_{t=0} = \operatorname{div}_M(\dot{x}^{tg}) - 2H\langle \dot{x}, N\rangle.$$

Thus,
$$\left.\frac{d}{dt}\operatorname{Area}(M_t)\right|_{t=0} = \int_M \operatorname{div}_M(\dot{x}^{tg}) - 2\int_M H\langle \dot{x}, N\rangle$$

and applying the divergence theorem finishes the proof. □

Remark 1.6. Suppose $\dot{x}_t = HN$, then
$$\left.\frac{d}{dt}(\operatorname{Area}(M_t))\right|_{t=0} = -2\int_M H^2 dM,$$

so the area is decreasing (respectively, increasing) if we deform the surface in the direction (opposite direction) of the so-called mean curvature vector HN.

Lemma 1.7. *Let $x : M \to \mathbb{R}^3$ be an isometric immersion. The surface M is a minimal surface if and only if the first variation of the area of M along any variation is zero, i.e., $x : M \to \mathbb{R}^3$ is a critical point for the area functional.*

Proof. By the first variation formula, if $H \equiv 0$ then clearly M is a critical point of the area functional for variations fixing the boundary. Suppose that $H \not\equiv 0$ and let $p \in M$ be a point where $H(p) \neq 0$. Let U be a neighbourhood of p in M where H is never zero and let $\phi \colon M \to \mathbb{R}$ be a non-negative smooth function compactly supported in U. Then, by considering the variation $x_t := x + \phi N$ in equation (1.1) one obtains
$$\frac{d}{dt}\operatorname{Area}(M_t)|_{t=0} = -2\int_M H\phi dM = -2\int_U H\phi dM \neq 0.$$

In other words, if the mean curvature is non-zero at a point, there exists a variation of M such that $\frac{d}{dt}\operatorname{Area}(M_t)|_{t=0} \neq 0$ and thus M is not a critical point for the area functional. This finishes the proof of the lemma. □

In light of Lemma 1.7, we are able to give an equivalent definition of minimal surfaces.

Definition 1.8 (Second definition). An isometric immersion $x : M \to \mathbb{R}^3$ is a *minimal* isometric immersion if it is a critical point for the area functional.

Remark 1.9. Note that the second definition is the more geometric one and in this sense it is a more natural definition.

Definition 1.10. A compact minimal surface M with boundary ∂M is absolute area minimizing if $\text{Area}(M) \leq \text{Area}(\Sigma)$ for any other surface Σ with $\partial \Sigma = \partial M$.

Since a minimum is always obtained at a critical point we have the following corollary.

Corollary 1.11. *An absolute area minimizing surface is a minimal surface.*

1.3. Convex hull and other properties

In this section we discuss a few geometric properties of minimal surfaces.

Lemma 1.12. *Let $x : M \to \mathbb{R}^3$ be an isometric immersion and let $V \in \mathbb{R}^3$ be a constant vector. Then*

$$\Delta_M \langle x, V \rangle = 2H \langle N, V \rangle$$

and

$$\Delta_M \langle N, V \rangle = -|A|^2 \langle N, V \rangle - 2 \langle \nabla H, V \rangle.$$

Proof. Let e_i, $i = 1, 2$, be a geodesic frame. Then

$$\Delta_M \langle x, V \rangle = \sum_i e_i e_i \langle x, V \rangle = e_i \langle e_i, V \rangle = \sum_i \langle e_i(e_i), V \rangle$$

$$= \sum_i \langle \nabla_{e_i} e_i + \langle Ae_i, e_i \rangle N, V \rangle = 2H \langle N, V \rangle$$

and

$$\Delta_M \langle N, V \rangle = \sum_i e_i e_i \langle N, V \rangle = \sum_i e_i \langle -Ae_i, V \rangle = \sum_i \langle e_i(-Ae_i), V \rangle$$

$$= -\sum_i \langle \nabla_{e_i}(Ae_i) + \langle Ae_i, Ae_i \rangle N, V \rangle$$

$$= -|A|^2 \langle N, V \rangle - \sum_{i,j} \langle \nabla_{e_i}(\langle Ae_i, e_j \rangle e_j), V \rangle$$

$$= -|A|^2 \langle N, V \rangle - \sum_{i,j} \langle \nabla_{e_i}(\langle Ae_j, e_i \rangle e_j), V \rangle$$

$$= -|A|^2 \langle N, V \rangle - \sum_{i,j} \langle \langle \nabla_{e_i}(Ae_j), e_i \rangle e_j, V \rangle$$

$$= -|A|^2 \langle N, V \rangle - \sum_{i,j} \langle \langle \nabla_{e_j}(Ae_i), e_i \rangle e_j, V \rangle$$

$$= -|A|^2 \langle N, V \rangle - \sum_{i,j} \langle e_j(\langle Ae_i, e_i \rangle) e_j, V \rangle$$

$$= -|A|^2 \langle N, V \rangle - 2 \langle \nabla H, V \rangle. \qquad \square$$

In light of Lemma 1.12 we give a third definition for minimal surfaces.

Definition 1.13 (Third definition). An isometric immersion $x : M \to \mathbb{R}^3$ is a *minimal* isometric immersion if the coordinate functions are harmonic functions.

Remark 1.14. Taking $H = 0$ and $V = (0, 0, 1)$ in Lemma 1.12 we observe that the function $f = N_3$ satisfies $\Delta_M f + |A|^2 f = 0$. Such a function is called a Jacobi function as we will later, in Section 4.1, see.

Definition 1.15. The convex hull $\mathcal{E}(\Omega)$ of a set $\Omega \subset \mathbb{R}^3$ is defined as the smallest convex set of $\mathbb{R}^3$ that contains Ω. We remark that if Ω is compact, then $\mathcal{E}(\Omega)$ is equal to the intersection of all half spaces Π that contain Ω.

Proposition 1.16 (Convex hull property for minimal surfaces). *If $\Sigma \subset \mathbb{R}^3$ is a compact minimal surface, then $\Sigma \subset \mathcal{E}(\partial \Sigma)$.*

Proof. Let $\Pi \subset \mathbb{R}^3$ be a half space. Then Π can be written as

$$\Pi = \{x \in \mathbb{R}^3 : x \cdot e \leq a\}$$

for a constant vector e and some constant $a \in \mathbb{R}$. Since the coordinate functions on Σ are harmonic, so is the function $u(x) = x \cdot e$. Hence, by the maximum principle for harmonic functions, u attains its maximum on $\partial \Sigma$ and therefore there exists $x_0 \in \partial \Sigma$ such that, for all $x \in \Sigma$, $u(x) \leq u(x_0)$. Thus, if $\partial \Sigma \subset \Pi$ then, for all $x \in \Sigma$, $u(x) \leq u(x_0) \leq a$, which means that Σ is contained in Π. Therefore, $\Sigma \subset \mathcal{E}(\partial \Sigma)$. $\qquad \square$

Corollary 1.17. *If $\Sigma \subset \mathbb{R}^3$ is a compact minimal disc and K is a compact, convex set such that $K \cap \partial \Sigma = \emptyset$, then $K \cap \Sigma$ is simply connected (i.e., any closed curve $\gamma \subset K \cap \Sigma$ bounds a disc in $K \cap \Sigma$).*

Proof. Let $\gamma \subset K \cap \Sigma$ be a closed curve that does not bound a disc in $K \cap \Sigma$. Since Σ is a disc, γ bounds a disc D in Σ, which is not contained in K. This violates the convex hull property, since $\partial D = \gamma \subset K$. This contradiction proves the corollary. $\square$

1.4. Minimal graphs

Let Ω be a domain in $\mathbb{R}^2$, let
$$u \colon \Omega \to \mathbb{R}$$
be a smooth function and consider the surface
$$M := (x, y, u(x,y)) \subset \mathbb{R}^3, \quad (x,y) \in \Omega.$$
The Gauss map of M is given by
$$N = \frac{1}{\sqrt{1 + |Du|^2}}(-u_x, -u_y, 1)$$
(upward pointing unit normal). It can be checked that
$$2H = \operatorname{div} \frac{Du}{\sqrt{1 + |Du|^2}}$$
$$= \frac{1}{(1 + |Du|^2)^{\frac{3}{2}}} \left((1 + u_y^2) u_{xx} - 2 u_x u_y u_{xy} + (1 + u_x^2) u_{yy} \right).$$

Definition 1.18. The graph of a function u is a minimal graph when u satisfies the following equation
$$\operatorname{div}\left(\frac{Du}{\sqrt{1 + |Du|^2}} \right) = 0, \tag{1.2}$$
i.e., the graph of u is a minimal surface.

Equation (1.2) is called the minimal surface equation and it is the prototypical quasi-linear elliptic partial differential equation.

2. The Bernstein Theorem

Theorem 2.1 (Bernstein theorem [2]). *Let $u \colon \mathbb{R}^2 \to \mathbb{R}$ be a minimal graph. Then $(x, y, u(x,y))$ is a flat plane. Namely, the graph is given by $(x, y, ax + by + c)$, $a, b, c \in \mathbb{R}$.*

Remark 2.2. From the minimal graph equation it is clear that there is a strong relation between minimal graphs and harmonic functions, especially when $|Du|$ is small. In fact, many of the properties that harmonic functions satisfy are also satisfied by minimal graphs.

Since there exist many entire harmonic functions that are not of the form $ax + by + c$, $a, b, c \in \mathbb{R}$, e.g., $x^2 - y^2$, the Bernstein theorem is a key result that, among other things, illustrates how harmonic functions and minimal graphs are different.

Remark 2.3. The Bernstein problem refers to the same question but in higher dimension. Namely, given $u \colon \mathbb{R}^{n-1} \to \mathbb{R}$ a minimal graph, is the graph of u a flat plane?

It turns out that the answer is yes for $n \leq 8$; this was proven by De Giorgi for $n = 4$ in [6], Almgren, Jr. for $n = 5$ in [1] and Simons for $n \leq 8$ in [14]. For $n > 8$ the answer is no, as the examples constructed by Bombieri–De Giorgi–Giusti in [3] show.

The proof of the Bernstein theorem we present here, which follows [13], is far from being Bernstein's original proof, which uses techniques from complex analysis.

2.1. *Proof of the Bernstein theorem*

Let $u \colon \mathbb{R}^2 \to \mathbb{R}$ be a minimal graph and let $M := \text{graph}(u)$. To prove the Bernstein theorem, we are going to show that

$$\int_M |A|^2 = 0.$$

After possibly applying a vertical translation we are going to assume that $\vec{0} \in M$. Let $B(R)$ denote the Euclidean ball of radius R centred at the origin.

Lemma 2.4 (Quadratic area growth). *For any $R > 0$,*

$$\text{Area}(B(R) \cap M) \leq \tfrac{1}{2} \text{Area}(\partial B(R)) = 2\pi R^2.$$

Before proving this lemma, we are going to show that a minimal graph $u \colon \Omega \to \mathbb{R}$ is area minimizing in $\Omega \times \mathbb{R}$. Namely, we have the following claim.

Claim 2.5. *Let Σ be a surface in $\Omega \times \mathbb{R}$ with $\partial \Sigma = \partial \,\text{graph}(u)$. Then*

$$\text{Area}(\Sigma) \geq \text{Area}(\text{graph}(u)).$$

Proof. Let $V \colon \Omega \times \mathbb{R} \to \mathbb{R}^3$ be the vector field defined by

$$V(x,y,z) := N(x,y) = \frac{1}{\sqrt{1+|\nabla u|^2}}(-u_x, -u_y, 1).$$

Let Δ be the "region" between Σ and graph(u) and suppose first that the upward pointing unit normal vector N to graph(u) is pointing into Δ. Then

$$\operatorname{div}_{\mathbb{R}^3} V = \operatorname{div}_{\mathbb{R}^2}\left(\frac{1}{\sqrt{1+|Du|^2}}(-u_x, -u_y)\right) + \partial_z\left(\frac{1}{\sqrt{1+|Du|^2}}\right) = 0.$$

Note that the ∂_z-term is zero because $\frac{1}{\sqrt{1+|Du|^2}}$ does not depend on z and the $\operatorname{div}_{\mathbb{R}^2}$-term is equal to $-\operatorname{div}\left(\frac{Du}{\sqrt{1+|Du|^2}}\right)$, which is zero because we are working with a minimal graph. Applying the divergence theorem (Stokes theorem) to Δ yields

$$0 = \int_\Delta \operatorname{div}_{\mathbb{R}^3} V = \int_{\partial \Omega} V \cdot \xi,$$

where ξ is the outward pointing unit normal to $\partial \Delta$. Note that

$$\partial \Delta = \operatorname{graph}(u) \cup \Sigma.$$

Therefore,

$$\begin{aligned} 0 &= \int_{\partial \Omega} V \cdot \xi \\ &= \int_{\operatorname{graph}(u)} V \cdot (-N) + \int_\Sigma V \cdot \xi = -\operatorname{Area}(\operatorname{graph}(u)) + \int_\Sigma V \cdot \xi, \end{aligned}$$

which gives

$$\operatorname{Area}(\operatorname{graph}(u)) = \int_\Sigma V \cdot \xi \leq \int_\Sigma |V \cdot \xi| \leq \int_\Sigma |V||\xi| \leq \operatorname{Area}(\Sigma).$$

In the general case, when N might not be pointing into Δ, we apply the above argument to each connected component of Δ, replacing N with $-N$ when N does not point into the component. $\square$

Remark 2.6. If Ω is convex, by the convex hull property, a compact minimal surface Σ with $\partial \Sigma \subset \partial \Omega \times \mathbb{R}$ must be contained in $\Omega \times \mathbb{R}$. Therefore, if Ω is convex, by Claim 2.5 and Corollary 1.11, a minimal graph over Ω is an absolute area minimizer.

Note, however, that when Ω is not convex, minimal graphs are not necessarily absolute area minimizing.

Proof of Lemma 2.4. In order to prove that
$$\text{Area}(B(R) \cap M) \leq \tfrac{1}{2}\text{Area}(\partial B(R)) = 2\pi R^2,$$
we are going to use the same arguments as in the proof of Claim 2.5.

Let $D(R)$ denote the disc in $\mathbb{R}^2$ of radius R centred at the origin and let $V: D(R) \times \mathbb{R} \to \mathbb{R}^3$ be the vector field defined by
$$V(x,y,z) := N(x,y) = \frac{1}{\sqrt{1+|\nabla u|^2}}(-u_x, -u_y, 1).$$

Let Δ be a region (not necessarily connected) between $M \cap B(R)$ and $\partial B(R)$ so that $\text{graph}(u) \subset \partial \Delta$. Note that given Δ, $\partial \Delta$ is equal to the union of $B(R) \cap M$ and a region S of the sphere. Thus, by possibly changing the region Δ, we can assume that
$$\text{Area}(S) \leq \tfrac{1}{2}\text{Area}(\partial B(R)) = 2\pi R^2.$$

Then, the area bound follows by applying the same arguments as in the proof of Claim 2.5 with Σ replaced by S. □

Lemma 2.7. Let $M \subset \mathbb{R}^3$ be a minimal graph. Then, for any $\phi \in C_0^\infty(M)$,
$$\int_M |A|^2 \phi^2 \leq \int_M |\nabla \phi|^2.$$

Proof. Since M is a graph, the function $f = N_3$, that is the third component of the normal, is a positive function and by Remark 1.14 it satisfies $\Delta_M f + |A|^2 f = 0$. Since f is positive, the function $w = \log f$ is well defined and it satisfies
$$\Delta_M w = \text{div}_M(\nabla f)$$
$$= \text{div}_M\left(\frac{\nabla f}{f}\right) = \frac{1}{f}\Delta_M f - \frac{1}{f^2}|\nabla f|^2 = -|A|^2 - |\nabla w|^2.$$

Now let $\phi \in C_0^\infty(M)$. Using integration by parts and the Cauchy–Schwarz inequality we obtain
$$\int_M \phi^2 |A|^2 + \int_M \phi^2 |\nabla w|^2 = -\int_M \phi^2 \Delta_M w = \int_M \nabla \phi^2 \cdot \nabla w$$
$$\leq 2\int_M |\phi||\nabla \phi||\nabla w|$$
$$\leq \int_M \phi^2 |\nabla w|^2 + \int_M |\nabla \phi|^2.$$

Cancelling the $\int_M \phi^2 |\nabla w|^2$ term on both sides yields the result. □

Remark 2.8. In order to obtain the conclusion of Lemma 2.7, the minimal surface M does not have to be a minimal graph. This lemma holds as long as there exists $f\colon M \to \mathbb{R}$ that is a positive solution of $\Delta_M f + |A|^2 f = 0$.

Proof of Theorem 2.1. Given $x \in M$ let $r(x) := |x|$ denote the Euclidean distance between x and the origin. As mentioned before, we will show that

$$\int_M |A|^2 = 0.$$

We will do this by using Lemma 2.7 with an appropriate choice of ϕ and this procedure is known as the "logarithmic cut-off trick." Let $\phi_R \colon M \to \mathbb{R}$ be defined as follows:

$$\phi_R(x) := \begin{cases} 1 & \text{if } r^2(x) \le R, \\ 2 - \dfrac{\ln(r^2(x))}{\ln R} & \text{if } R \le r^2(x) \le R^2, \\ 0 & \text{if } r^2(x) \ge R^2. \end{cases}$$

Note that ϕ is not smooth. However, it is a Lipschitz function and by a standard approximation argument it is easy to check that Lemma 2.7 holds for Lipschitz functions as well. Note that

$$|\nabla \phi_R|^2 \le \frac{4}{r^2 (\ln R)^2}.$$

Given $t > 0$, let $M(t) := M \cap B(t)$. Fix $N := [\ln \sqrt{R}]$, the integer part of $\ln \sqrt{R}$. Then

$$R < \sqrt{R} e^{N+1} < R + e\sqrt{R} \implies N + 1 < \ln \frac{R + e\sqrt{R}}{\sqrt{R}}.$$

Using Lemmas 2.7 and 2.4, we have that

$$\int_{M(\sqrt{R})} |A|^2 \le \int_M |A|^2 \phi_R^2 \le \int_{M(R)} |\nabla \phi_R|^2$$

$$\le \int_{M(R) - M(\sqrt{R})} \frac{4}{r^2 (\ln R)^2}$$

$$\le \frac{4}{(\ln R)^2} \sum_{k=0}^{N} \int_{M(\sqrt{R} e^{k+1}) - M(\sqrt{R} e^k)} \frac{1}{r^2}$$

$$\leq \frac{4}{(\ln R)^2} \sum_{k=0}^{N} \frac{1}{(\sqrt{R}e^k)^2} \operatorname{Area}(M(\sqrt{R}e^{k+1}))$$

$$\leq \frac{4}{(\ln R)^2} \sum_{k=0}^{N} \frac{2\pi R e^{2k+2}}{Re^{2k}} = \frac{8\pi}{(\ln R)^2} \sum_{k=0}^{N} e^2$$

$$= \frac{8\pi e^2 (N+1)}{(\ln R)^2} \leq 8\pi e^2 \frac{\ln \frac{R+e\sqrt{R}}{\sqrt{R}}}{(\ln R)^2}.$$

Since

$$\lim_{R \to \infty} \frac{\ln \frac{R+e\sqrt{R}}{\sqrt{R}}}{(\ln R)^2} = 0,$$

this finishes the proof of the Bernstein theorem. □

3. Quadratic Area Growth and Parabolicity

In this section we prove that a complete surface with intrinsic quadratic area growth is parabolic (in the function-theoretical sense). The proof is very similar to the proof of the Bernstein theorem which can in fact be seen as a corollary of this result.

Theorem 3.1. *Let* $\Sigma \subset \mathbb{R}^3$ *be a complete surface with* $\vec{0} \in \Sigma$ *and quadratic area growth, i.e., for any* $s > 0$ *we have that*

$$\operatorname{Area}(B_s^\Sigma) \leq Cs^2,$$

where $B_s^\Sigma(x)$ *denotes the geodesic ball of radius* s *centred at* $x \in \Sigma$ *and* $B_s^\Sigma := B_s^\Sigma(\vec{0})$. *Then* Σ *is parabolic, i.e., any positive super harmonic function* u *on* Σ ($\Delta_\Sigma u \leq 0$) *is constant.*

Remark 3.2. The universal cover of a complete surface with quadratic area growth is conformal to $\mathbb{C}$.

Remark 3.3. The converse of the theorem is not true. The helicoid is parabolic but it does not have quadratic area growth.

Remark 3.4. The Bernstein theorem can be seen as a corollary of Theorem 3.1.

Recall that if M is an entire minimal graph, then Lemma 2.4 gives that Area$(\Sigma \cap B(s)) \leq 2\pi s^2$. This implies that M has intrinsic quadratic area growth because

$$B_s^\Sigma \subset B(s) \cap M \implies \text{Area}(B_s^\Sigma) \leq \text{Area}(\Sigma \cap B(s)).$$

Therefore, by Theorem 3.1, M is parabolic.

Since $\Delta_M N_3 = -|A|^2 N_3$ and M is an entire minimal graph, then N_3 is a positive super harmonic function. Applying Theorem 3.1 then gives that N_3 is a positive constant. Therefore $|A|^2 N_3 = 0$ which implies $|A|^2 = 0$.

Proof of Theorem 3.1. Indeed, the proof is similar to the proof of the Bernstein theorem. Let $u > 0$ be such that $\Delta_\Sigma u \leq 0$ and set $w = \log u$. Then

$$\Delta_\Sigma w = \text{div}_\Sigma \left(\frac{|\nabla u|}{u} \right)$$

$$= \frac{1}{u} \Delta_\Sigma u - \frac{1}{u^2} |\nabla u|^2 \leq -|\nabla w|^2 \implies |\nabla w|^2 \leq -\Delta_\Sigma w.$$

We will show that w is constant. For a positive Lipschitz function η with compact support we have

$$\int_\Sigma \eta^2 |\nabla w|^2 \leq -\int_\Sigma \eta^2 \Delta_\Sigma w = \int_\Sigma \nabla \eta^2 \cdot \nabla w$$

$$\leq 2 \int_\Sigma \eta |\nabla \eta| |\nabla w|$$

$$\leq \frac{1}{2} \int_\Sigma \eta^2 |\nabla w|^2 + 2 \int_\Sigma |\nabla \eta|^2,$$

where we have used integration by parts and the inequality

$$2ab = 2 \frac{a}{\sqrt{2}} (\sqrt{2} b) \leq \frac{1}{2} a^2 + 2b^2.$$

The above inequality gives

$$\int_\Sigma \eta^2 |\nabla w|^2 \leq 4 \int_\Sigma |\nabla \eta|^2.$$

The proof that w is constant, namely that $|\nabla w| = 0$ now follows using the same arguments as in the proof of Bernstein theorem, using now $r(x)$ as the intrinsic distance and the logarithmic cut-off trick. □

4. Stable Minimal Surfaces

In this section we prove an important generalization of the Bernstein theorem.

Definition 4.1. Let $x\colon M \to \mathbb{R}^3$ be a minimal immersion. A minimal surface is stable if for any smooth variation $x_t\colon M \times (-\epsilon, \epsilon) \to \mathbb{R}^3$ of $x\colon M \to \mathbb{R}^3$ we have
$$\left.\frac{d^2}{dt^2} \operatorname{Area}(M_t)\right|_{t=0} \geq 0.$$

Remark 4.2. As in Corollary 1.11, we observe that any absolute area minimizing surface is a stable minimal surface.

Theorem 4.3. *Let M be a complete stable minimal surface. Then M is a flat plane.*

In these notes we are assuming that M is orientable and, in this case, Theorem 4.3 was proved independently by Fischer-Colbrie and Schoen [8], do Carmo and Peng [7], and Pogorelov [10]. Later, Ros [11] proved that a complete, non-orientable minimal surface in $\mathbb{R}^3$ is never stable.

4.1. Second variation formula

Theorem 4.4 (Second variation formula). *Let $x\colon M \to \mathbb{R}^3$ be a minimal immersion. Given $\phi \in C_0^\infty(M)$, let $x_t\colon M \times (-\epsilon, \epsilon) \to \mathbb{R}^3$ be a smooth variation such that $\phi = \dot{x} \cdot N$, where $\dot{x} := \frac{dx_t}{dt}\big|_{t=0}$. Then*
$$\left.\frac{d^2}{dt^2} \operatorname{Area}(M_t)\right|_{t=0} = -\int_M \phi L(\phi),$$

where $L(\phi) = \Delta_M \phi + |A|^2 \phi$. Therefore, M is stable if and only if for any $\phi \in C_0^\infty(M)$, we have
$$\int_M |A|^2 \phi^2 \leq \int_M |\nabla \phi|^2. \tag{4.1}$$

Remark 4.5. Lemma 2.7, together with Theorem 4.4, implies that minimal graphs are stable. Therefore, Theorem 4.3 is a generalization of the Bernstein theorem.

Remark 4.6. A function u such that $Lu = 0$ is called a Jacobi function (and uN is called a Jacobi field). If there exists a positive Jacobi function u on M, then for any $\phi \in C_0^\infty(M)$ we have that

$$\int_M |A|^2 \phi^2 \leq \int_M |\nabla \phi|^2.$$

The proof of this fact is exactly the same as the proof of Lemma 2.7 using u instead of N_3. Then, by Theorem 4.4, if there exists a positive Jacobi function u on M then M is stable.

Remark 4.7. Theorem 4.3 does not hold for hyper-surfaces in $\mathbb{R}^n$, $n \geq 8$, since the Bernstein theorem does not hold. It is still an open problem in the remaining dimensions.

Proof of Theorem 4.4. Let $x_t \colon M \times (-\epsilon, \epsilon) \to \mathbb{R}^3$ be a smooth variation such that $\phi = \dot{x} \cdot N$. From the first variation formula we have

$$\frac{d}{dt} \operatorname{Area}(M_t) = -2 \int_{M_t} H_t \langle \dot{x}_t, N_t \rangle = -2 \int_M H_t \langle \dot{x}_t, N_t \rangle \frac{\sqrt{\det(g_{ij}(t))}}{\sqrt{\det(g_{ij}(0))}}.$$

Thus,

$$\frac{d^2}{dt^2} \operatorname{Area}(M_t) = -2 \int_M \left(\frac{d}{dt} H_t\right) \langle \dot{x}_t, N_t \rangle \frac{\sqrt{\det(g_{ij}(t))}}{\sqrt{\det(g_{ij}(0))}}$$

$$- 2 \int_M H_t \frac{d}{dt} \left(\langle \dot{x}_t, N_t \rangle \frac{\sqrt{\det(g_{ij}(t))}}{\sqrt{\det(g_{ij}(0))}}\right).$$

Since $H_0 = 0$, the second term will be zero when we substitute $t = 0$ and

$$\left.\frac{d^2}{dt^2} \operatorname{Area}(M_t)\right|_{t=0} = -2 \int_M \left.\frac{dH_t}{dt}\right|_{t=0} \langle \dot{x}, N \rangle = -2 \int_M \left.\frac{dH_t}{dt}\right|_{t=0} \phi$$

It thus remains to show that

$$2 \left.\frac{dH_t}{dt}\right|_{t=0} = \Delta_M \phi + |A|^2 \phi.$$

This is the result contained in the following claim.

Claim 4.8. Let $x \colon M \to \mathbb{R}^3$ be a minimal immersion and let $x_t \colon M \times (-\epsilon, \epsilon) \to \mathbb{R}^3$ be a smooth variation (not necessarily with compact support and fixing the boundary) such that $\phi = \dot{x} \cdot N$, then

$$2 \left.\frac{dH_t}{dt}\right|_{t=0} = \Delta_M \phi + \phi |A|^2.$$

Remark 4.9. Let $x_t\colon M \times (-\epsilon, \epsilon) \to \mathbb{R}^3$ be a variation (not necessarily with compact support and fixing the boundary) of a minimal immersion $x\colon M \to \mathbb{R}^3$ such that $\frac{dH_t}{dt}\big|_{t=0} = 0$ and let $u = \dot{x} \cdot N$. Then, by Claim 4.8 we have that

$$0 = \frac{dH_t}{dt}\bigg|_{t=0} = \Delta_M u + |A|^2 u.$$

Therefore u is a Jacobi function. This is a good way to obtain Jacobi functions.

For instance, let $I_t\colon \mathbb{R}^3 \times (-\epsilon, \epsilon) \to \mathbb{R}^3$ be a one-parameter family of isometries of $\mathbb{R}^3$ such that I_0 is the identity and let $x_t\colon M \times (-\epsilon, \epsilon) \to \mathbb{R}^3$ be the one-parameter deformation of $x\colon M \to \mathbb{R}^3$ defined by $x_t := I_t \circ x$. Then, since an isometry leaves the mean curvature invariant, by the previous observation, $\frac{dx_t}{dt}\big|_{t=0} \cdot N$ is a Jacobi function.

Letting $I_t(p) = p + te_i$, $i = 1, 2, 3$, in the previous discussion gives that $N_i = \langle N, e_i \rangle$, $i = 1, 2, 3$, are Jacobi functions.

Proof of Claim 4.8. Let $x\colon M \to \mathbb{R}^3$ be an isometric immersion and consider the deformation $x_t\colon M \to \mathbb{R}^3$. Taylor expansion yields

$$x_t = x + t\phi N + O(t^2).$$

Since we want to compute the derivative of the mean curvature of x_t at $t = 0$, as we have done in the proof of Theorem 1.3, we will ignore in the calculation any term where a t^2-term appears. Let e_i be a geodesic frame on M at a point p and we will compute the derivative of the mean curvature of M_t at p and $t = 0$. In order to compute the mean curvature H_t of M_t, we need to compute the first and second fundamental form, $g_{ij}(t)$ and $A_{ij}(t)$, of M_t. A tangent frame to M_t around p is given by

$$f_i = e_i x_t = e_i + te_i(\phi N) = e_i + t(\phi_i N - \phi A e_i).$$

Thus, the metric $g_{ij}(t)$ of M_t is given by

$$g_{ij}(t) = \langle f_i, f_j \rangle = \delta_{ij} - 2t\phi A_{ij}.$$

In order to compute the unit normal, N_t, to M_t we first compute

$$f_1 \times f_2 = N - t\phi_i e_i$$

and since $|f_1 \times f_2| = 1$ we have that

$$N_t = N - t\phi_i e_i.$$

The second fundamental form $A_{ij}(t)$ of M_t is given by
$$A_{ij}(t) = \langle e_j(f_i), N_t \rangle.$$
Since
$$e_j(f_i) = \langle Ae_i, e_j \rangle N + t(\phi_{ij} - \phi \langle Ae_i, Ae_j \rangle)N$$
$$+ t(-\phi_i Ae_j - \phi_j Ae_i - \phi \nabla_{e_j} Ae_i),$$
we have that
$$A_{ij}(t) = A_{ij} + t(\phi_{ij} - \phi A_{ij}^2).$$
Let M^* denote the adjoint of a matrix M; then
$$g_{ij}^{-1}(t) = \frac{1}{\det(g_{ij}(t))} g_{ij}^*(t) = \frac{1}{\det(g_{ij}(t))} (\delta_{ij} - 2t\phi A_{ij}^*).$$
Using the formula $\det(I + tM) = I + t\,\mathrm{Trace}(M) + O(t^2)$ we obtain
$$\det(g_{ij}(t)) = 1$$
and
$$g_{ij}^{-1}(t) = \delta_{ij} - 2t\phi A_{ij}^*.$$
Finally, the product of the matrix $g_{ij}^{-1}(t)$ with the matrix $A_{ij}(t)$ is
$$g_{ij}^{-1}(t) A_{ij}(t) = (\delta_{ij} - 2t\phi A_{ij}^*)(A_{ij} + t(\phi_{ij} - \phi A_{ij}^2))$$
$$= t(\phi_{ij} - \phi A_{ij}^2 - 2\phi A_{ij}^* A_{ij})$$
$$= t(\phi_{ij} - \phi A_{ij}^2 - 2\phi \det(A_{ij}) \delta_{ij}).$$
Using the fact that for a minimal surface in $\mathbb{R}^3$ we have that $|A|^2 = -2K$, where K denotes the Gaussian curvature, we obtain that the mean curvature of M_t at $x_t(p)$ is
$$2H(t) = \mathrm{trace}(g_{ij}^{-1}(t) A_{ij}(t))$$
$$= t(\Delta_M \phi - \phi|A|^2 - 4\phi \det(A_{ij}))$$
$$= t(\Delta_M \phi - \phi|A|^2 - 4\phi K)$$
$$= t(\Delta_M \phi + \phi|A|^2).$$
This yields
$$2\frac{dH_t}{dt}\bigg|_{t=0} = \Delta_M \phi + \phi|A|^2$$
and finishes the proof of the claim. □

This finishes the proof of Theorem 4.4. □

4.2. Quadratic area growth for stable minimal surfaces

In light of the proof of Theorem 3.1 and the second variation formula, Theorem 4.3 follows if we prove that a stable minimal surface has intrinsic quadratic area growth. This has been proven in [4].

Lemma 4.10. *Let Σ be a stable minimal surface in $\mathbb{R}^3$. If $B_s^\Sigma(x) \subset \Sigma$, $\partial B_s^\Sigma(x) \subset \Sigma \backslash \partial \Sigma$, then*

$$\mathrm{Area}(B_s^\Sigma(x)) \leq \tfrac{4}{3}\pi s^2.$$

Proof. We begin to show that it suffices to prove the result when the exponential map is a diffeomorphism on $B_s^\Sigma(x)$ and $B_r^\Sigma(x)$ is simply-connected for any $r \leq s$. To do this, we require the following lemma, the proof of which one can find for instance in [8].

Lemma 4.11. *A minimal surface Σ is stable if and only if Σ admits a positive Jacobi function.*

Recall Remark 4.6 for the definition of a Jacobi function and note that the "if" part of Lemma 4.11 is proven in that remark. Thanks to Lemma 4.11, it is rather straightforward to check that if Σ is a stable minimal surface then its universal cover $\widetilde{\Sigma}$ is also stable. In fact, if $u\colon \Sigma \to \mathbb{R}$ is a positive Jacobi function, then $\widetilde{u}\colon \widetilde{\Sigma} \to \mathbb{R}$ defined as $\widetilde{u}(\widetilde{x}) = u(x)$, where $\widetilde{x} \in \widetilde{\Sigma}$ is a lift of $x \in \Sigma$, is also a positive Jacobi function.

If the exponential map is not a diffeomorphism on $B_s^\Sigma(x)$, then let $\widetilde{\Sigma}$ denote the universal cover of Σ and let $\widetilde{x} \in \widetilde{\Sigma}$ be a lift of x. By the previous arguments, $B_s^{\widetilde{\Sigma}}(\widetilde{x})$ is also stable and we have that

$$\mathrm{Area}(B_s^\Sigma(x)) \leq \mathrm{Area}(B_s^{\widetilde{\Sigma}}(\widetilde{x})) \leq \tfrac{4}{3}\pi s^2.$$

Since the Gaussian curvature of a minimal surface immersed in $\mathbb{R}^3$ is non-positive, the exponential map is a diffeomorphism on $B_s^{\widetilde{\Sigma}}(\widetilde{x})$. Moreover, $B_r^{\widetilde{\Sigma}}(\widetilde{x})$ is simply-connected for any $r \leq s$. This shows that it suffices to prove the lemma when the exponential map is a diffeomorphism on $B_s^\Sigma(x)$ and $B_r^\Sigma(x)$ is simply-connected for any $r \leq s$.

Let $L(t)$ denote the length of $\partial B_t^\Sigma(x)$ for any $t \leq s$. Then, the first variation of arc length gives

$$\frac{d}{dt}L(t) = \int_{\partial B_t^\Sigma(x)} k_g,$$

where k_g is the geodesic curvature of $\partial B_t^\Sigma(x)$. Thus, using the Gauss–Bonnet theorem,

$$\frac{dL}{dt}(t) = \int_{\partial B_t^\Sigma(x)} k_g = 2\pi - \int_{B_t^\Sigma(x)} K_\Sigma, \tag{4.2}$$

where K_Σ denote the Gaussian curvature of Σ.

We recall now the following special case of the coarea formula (see for example [12]):

Let $f : \Sigma \to \mathbb{R}$ be a C^1 function; then

$$\int_\Sigma g|\nabla f| = \int_\mathbb{R} \int_{f^{-1}(s)} g,$$

for any smooth non-negative function $g : M \to \mathbb{R}$.

Integrating equation (4.2) twice and using the coarea formula (with f equal to the intrinsic distance to x), we obtain

$$\text{Area}(B_s^\Sigma(x)) - \pi s^2 = -\int_0^s \int_0^t \int_{B_\rho(x)} K_\Sigma.$$

Furthermore,

$$\int_0^s \int_0^t \int_{B_\rho(x)} K_\Sigma = \frac{1}{2} \int_0^s \frac{d^2}{dt^2}(s-t)^2 \left(\int_0^t \int_{B_\rho(x)} K_\Sigma \right)$$

$$= -\frac{1}{2} \int_0^s \frac{d}{dt}(s-t)^2 \left(\int_{B_t^\Sigma(x)} K_\Sigma \right)$$

$$= \int_0^s \int_{\partial B_t^\Sigma(x)} K_\Sigma \frac{(s-t)^2}{2}$$

and using again the coarea formula

$$\int_0^s \int_{\partial B_t^\Sigma(x)} K_\Sigma \frac{(s-t)^2}{2} = \int_{B_s^\Sigma(x)} K_\Sigma \frac{(s-r(y))^2}{2},$$

where for $y \in B_s^\Sigma(x)$, $r(y)$ denotes the intrinsic distance in Σ from x to y.

Recalling that for a minimal surface $-K_\Sigma = \frac{|A|^2}{2}$, so far we have obtained that

$$\text{Area}(B_s^\Sigma(x)) - \pi s^2 = \int_{B_s^\Sigma(x)} |A|^2 \frac{(s-r(y))^2}{4}. \tag{4.3}$$

We are now going to use the stability inequality, that is equation (4.1), to bound the right-hand side of equation (4.3). Using

$$\phi: B_s^\Sigma(x) \to \mathbb{R}, \quad \phi(y) = \frac{(s - r(y))}{2}$$

in the stability inequality, we obtain

$$\int_{B_s^\Sigma(x)} |A|^2 \frac{(s - r(y))^2}{4} \leq \int_{B_s^\Sigma(x)} \frac{1}{4} = \frac{\text{Area}(B_s^\Sigma(x))}{4}.$$

This, together with equation (4.3), gives that

$$\text{Area}(B_s^\Sigma(x)) - \pi s^2 = \int_{B_s^\Sigma(x)} |A|^2 \frac{(s - r(y))^2}{4} \leq \frac{\text{Area}(B_s^\Sigma(x))}{4}$$

and thus

$$\text{Area}(B_s^\Sigma(x)) \leq \tfrac{4}{3} \pi s^2. \qquad \square$$

Using Lemma 4.10 and the stability inequality, namely equation (4.1), the proof of Theorem 4.3 follows using arguments that are identical to the proof of Theorem 3.1.

5. Exercises

(1) A catenoid is given in parametric form by

$$x(u, v) = (a \cosh v \cos u, a \cosh v \sin u, av), \quad a \in \mathbb{R}_+.$$

Prove that a catenoid is a minimal surface.

(2) Prove that if M is a complete minimal surface of revolution, then M must be a plane or a catenoid.

(3) Prove that if M is a complete ruled minimal surface, then M must be a plane or a helicoid.

(4) Compute the second fundamental form of a surface that is given as the graph of a function.

(5) Prove that if a surface is given as the graph of a function u, then

$$2H = \text{div} \frac{Du}{\sqrt{1 + |Du|^2}}$$

$$= \frac{1}{(1 + |Du|^2)^{\frac{3}{2}}} \left((1 + u_y^2) u_{xx} - 2 u_x u_y u_{xy} + (1 + u_x^2) u_{yy} \right).$$

(6) Let Ω be a bounded smooth domain of $\mathbb{R}^2$ and let $u\colon \Omega \to \mathbb{R}$ be a smooth function, $u \in C^\infty(\Omega) \cap C^0(\overline{\Omega})$. Suppose that the mean curvature of $(x, y, u(x, y))$ is non-positive. Show that if u obtains its minimum at an interior point, then it is constant.
(7) Prove that the helicoid is conformal to $\mathbb{C}$.
(8) In light of Remark 4.9, compute the Jacobi functions induced by rotations and dilations.

Hints for selected exercises

(6) Let $M = \operatorname{graph}(u)$ and define $f(x) = \langle x, e_3 \rangle$. Recalling Lemma 1.12, we observe that the function f satisfies $\Delta_M f = 2H \langle N, e_3 \rangle \leq 0$. One can now use the strong maximum principle to obtain the result.
(7) It is easy to check that the following parametrization of a helicoid is conformal:

$$f(u,v)\colon \mathbb{R}^2 \to \mathbb{R}^3, \quad f(u,v) = (\sinh u \cos v, \sinh u \sin v, v).$$

(8) Let $x\colon M \to \mathbb{R}^3$ be a minimal surface. Consider the matrix

$$R_t = \begin{pmatrix} \cos t & -\sin t & 0 \\ \sin t & \cos t & 0 \\ 0 & 0 & 1 \end{pmatrix}$$

and define $I_t \colon \mathbb{R}^3 \times (-\varepsilon, \varepsilon) \to \mathbb{R}^3$ by $I_t(x) = R_t x$. For each t, I_t defines the counter-clockwise rotation through an angle t leaving the x_3-axis fixed and thus I_t is a one-parameter family of isometries of $\mathbb{R}^3$. The Jacobi function induced by this rotation is then given by $\frac{dx_t}{dt}\big|_{t=0} \cdot N$, where $x_t \colon I_t \circ x$. Since

$$\frac{dR_t}{dt}\bigg|_{t=0} = \begin{pmatrix} 0 & -1 & 0 \\ 1 & 0 & 0 \\ 0 & 0 & 1 \end{pmatrix} =: B_0,$$

the function $B_0 x \cdot N$ is a Jacobi function. Using similar arguments, it is possible to show that $Bx \cdot N$ is a Jacobi function for any rotation B.

One works similarly with the dilations, considering now the one-parameter family I_t given by $I_t(x) = (1+t)x$. Note that, even though dilations are not isometries, a minimal surface stays minimal under a dilation and so the derivative of the mean curvature at $t = 0$ is zero. In this case, the Jacobi function we obtain is given by $x \cdot N$.

References

[1] F. J. Almgren, Some interior regularity theorems for minimal surfaces and an extension of Bernstein's theorem. *Ann. of Math.* (2) **84**, 277–292 (1966).

[2] S. Bernstein, Uber ein geometrisches theorem und seine anwendung auf die partiellen differentialglechungen vom auf die partiellen differentialglechungen vom elliptischen typus. *Math. Z.* **26**, 551–558 (1927). MR1544873, Zbl JFM 53.0670.01.

[3] E. Bombieri, E. De Giorgi and E. Giusti (Pisa), Minimal cones and the Bernstein problem. *Invent. Math.* **7**, 243–268 (1969). MR0250205,Zbl 0183.25901.

[4] T. H. Colding and W. P. Minicozzi II, Estimates for parametric elliptic integrands. *Int. Math. Res. Notices*, **6**, 291–297 (2002). MR1877004 (2002k:53060), Zbl 1002.53035.

[5] T. H. Colding and W. P. Minicozzi II, *A Course in Minimal Surfaces*. Graduate Studies in Mathematics, American Mathematical Society (2011).

[6] E. De Giorgi, *Frontiere orientate di misura minima*. Seminario di Matematica della Scuola Normale Superiore di Pisa, 1960–1961. Editrice Tecnico Scientifica, Pisa (1961).

[7] M. do Carmo and C. K. Peng, Stable complete minimal surfaces in $\mathbb{R}^3$ are planes. *Bull. Amer. Math. Soc.*, **1**, 903–906 (1979). MR0546314, Zbl 442.53013.

[8] D. Fischer-Colbrie and R. Schoen, The structure of complete stable minimal surfaces in 3-manifolds of nonnegative scalar curvature. *Comm. Pure Appl. Math.* **33**, 199–211 (1980). MR0562550, Zbl 439.53060.

[9] R. Osserman, *A Survey of Minimal Surfaces*. 2nd edn. Dover Publications, New York (1986). MR0852409, Zbl 0209.52901.

[10] A. V. Pogorelov, On the stability of minimal surfaces. *Soviet Math. Dokl.* **24**, 274–276 (1981). MR0630142, Zbl 0495.53005.

[11] A. Ros, One-sided complete stable minimal surfaces. *J. Differential Geom.* **74**, 69–92 (2006). MR2260928, Zbl 1110.53009.

[12] L. Simon, Lectures on geometric measure theory. In *Proceedings of the Center for Mathematical Analysis*, Vol. 3, Australian National University, Canberra, Australia (1983). MR0756417, Zbl 546.49019.

[13] L. Simon, The minimal surface equation. In *Geometry V: Minimal Surfaces*, pp. 239–266. Springer, Berlin (1997).

[14] J. Simons, Minimal varieties in Riemannian manifolds. *Ann. of Math.* **88**, 62–105 (1968). MR0233295, Zbl 0181.49702.

Chapter 6

Syzygies and Minimal Resolutions

F. E. A. Johnson

Department of Mathematics, University College London
Gower Street, London WC1E 6BT, UK
feaj@math.ucl.ac.uk

The essence of linear algebra over a field resides in the fact that every vector space is free; that is, has a spanning set of linearly independent vectors. The study of linear algebra over more general rings attempts to approximate this situation by the method of free resolutions. When a module M is not free we make a first approximation to its being free by taking a surjective homomorphism $\epsilon : F_0 \to M$ where F_0 is free to obtain an exact sequence

$$0 \to J_1 \to F_0 \xrightarrow{\epsilon} M \to 0.$$

Repeating the construction we approximate J_1 in turn by a free module to obtain an exact sequence $0 \to J_2 \to F_1 \to J_1 \to 0$. Iterating and splicing we obtain a *free resolution of* M in the sense of Hilbert [Über die theorie der algebraischen formen. *Math. Ann.* **36**, 473–539 (1890)]

$$\xrightarrow{\partial_{n+1}} F_n \xrightarrow{\partial_n} F_{n-1} \xrightarrow{\partial_{n-1}} \cdots \xrightarrow{\partial_2}$$
$$\searrow \nearrow$$
$$J_n$$

$$F_1 \xrightarrow{\partial_1} F_0 \xrightarrow{\epsilon} M \to 0$$
$$\searrow \nearrow$$
$$J_1$$

We study the relationship between the intermediate modules J_n, the so-called *syzygies of* M, and those free resolutions of M which are in some sense minimal.

1. Introduction

The notion of *free resolution* has its origin in the classical theory of invariants [2] and the study of graded modules over polynomial rings $\mathbf{F}[x_1, \ldots, x_n]$ where $\mathbf{F}$ is a field. In this context there is a well-defined notion of *minimal free resolution*. Such minimal resolutions have a strong uniqueness property; not only are they themselves unique up to isomorphism but in addition any other free resolution is a direct sum of the minimal free resolution with a free acyclic complex. In [1], Eilenberg gave an extension of this uniqueness property by essentially formal arguments. However, despite the elegance of Eilenberg's approach, its scope remains relatively narrow.

The main technical limitation of Eilenberg's theory arises from his definition of minimality. This places so strong a restriction on the class of rings to which it may be applied as to render it *a priori* inapplicable to many cases of interest. Consequently, we are forced to reformulate matters in a rather more general context.

Our primary notion is that of a *special class* $\mathfrak{S}$ of projectives in an abelian category $\mathfrak{A}$; the precise formulation is given in Section 6. Suffice to say here that $\mathfrak{S}$ plays a role analogous to that of finitely generated stably free modules over a ring. For an object $M \in \mathfrak{A}$ we consider $\mathfrak{S}$-*resolutions of* M, that is, exact sequences in $\mathfrak{A}$ of the form

$$\mathbf{S} = (\cdots \stackrel{\partial_{n+1}}{\to} S_n \stackrel{\partial_n}{\to} \cdots \to S_1 \stackrel{\partial_1}{\to} S_0 \to M \to 0),$$

where each $S_r \in \mathfrak{S}$. To such an resolution we may add an $\mathfrak{S}$-resolution of 0

$$\mathbf{T} = (\cdots \to T_n \to \cdots \to T_1 \to T_0 \to 0)$$

to obtain another $\mathfrak{S}$-resolution $\mathbf{S} \oplus \mathbf{T}$ of M thus

$$\mathbf{S} \oplus \mathbf{T} = (\cdots \to S_n \oplus T_n \to \cdots \to S_1 \oplus T_1 \to S_0 \oplus T_0 \to M \to 0).$$

$\mathbf{S}$ is said to be *minimal* when, for any $\mathfrak{S}$-resolution $\mathbf{S}'$ there exists a commutative diagram

$$\begin{pmatrix} \cdots \stackrel{\partial'_{n+1}}{\to} & S'_n & \stackrel{\partial'_n}{\to} & \cdots \stackrel{\partial'_1}{\to} & S'_0 & \stackrel{\eta}{\to} & M & \to 0 \\ & \varphi_n \downarrow & & & \varphi_0 \downarrow & & \downarrow \mathrm{Id}_M & \\ \cdots \stackrel{\partial_{n+1}}{\to} & S_n & \stackrel{\partial_n}{\to} & \cdots \stackrel{\partial_1}{\to} & S_0 & \stackrel{\epsilon}{\to} & M & \to 0 \end{pmatrix},$$

where each φ_r is epimorphic. When they exist, minimal resolutions are unique in the following sense:

Theorem 1.1. *Let $\mathbf{S}$ and $\widetilde{\mathbf{S}}$ be $\mathfrak{S}$-resolutions of M. If $\mathbf{S}$ is minimal then $\widetilde{\mathbf{S}} \cong \mathbf{S} \oplus \mathbf{T}$ for some $\mathfrak{S}$-resolution $\mathbf{T}$ of 0. In particular, if $\widetilde{\mathbf{S}}$ is also minimal then $\widetilde{\mathbf{S}} \cong \mathbf{S}$.*

In applications the requirement that M has an $\mathfrak{S}$-resolution is usually a very strong restriction. We may relax it by considering partial $\mathfrak{S}$-resolutions or *n-stems*. Thus an *n-stem* over M is an exact sequence in $\mathfrak{A}$ of the form

$$\mathbf{S} = (S_n \xrightarrow{\partial_n} \cdots \to S_1 \xrightarrow{\partial_1} S_0 \to M \to 0)$$

where each $S_r \in \mathfrak{S}$. The *n*-stem $\mathbf{S}^{(n)}$ is *minimal* when, for any *n*-stem $\widetilde{\mathbf{S}}^{(n)}$ there exists a commutative diagram

$$\begin{pmatrix} \widetilde{S}_n & \xrightarrow{\widetilde{\partial}_n} & \cdots & \xrightarrow{\widetilde{\partial}_1} & \widetilde{S}_0 & \xrightarrow{\widetilde{\eta}} & M & \to 0 \\ \varphi_n \downarrow & & & & \varphi_0 \downarrow & & \downarrow \mathrm{Id}_M & \\ S_n & \xrightarrow{\partial_n} & \cdots & \xrightarrow{\partial_1} & S_0 & \xrightarrow{\epsilon} & M & \to 0 \end{pmatrix}$$

in which each φ_r is epimorphic. For *n*-stems, Theorem 1.1 is modified as follows.

Theorem 1.2. *If $\mathbf{S}^{(n)}, \widetilde{\mathbf{S}}^{(n)}$ are n-stems over M and $\mathbf{S}^{(n)}$ is minimal, then $\widetilde{\mathbf{S}}^{(n-1)} \cong_{\mathrm{Id}_m} \mathbf{S}^{(n-1)} \oplus \mathbf{T}^{(n-1)}$ for some $(n-1)$-stem $\mathbf{T}^{(n-1)}$ over 0.*

If $\mathbf{S}^{(n)} = (S_n \xrightarrow{\partial_n} \cdots \to S_1 \xrightarrow{\partial_1} S_0 \to M \to 0)$ is an *n*-stem its *syzygies* are the intermediate objects $(J_r)_{1 \le r \le n}$ obtained via the canonical decomposition of ∂_r as the composition of a monomorphism i_r and an epimorphism p_r thus:

$$\begin{array}{ccc} S_r & \xrightarrow{\partial_r} & S_{r-1} \\ & p_r \searrow \quad \nearrow i_r & \\ & J_r & \end{array}$$

Minimality also implies a relation amongst syzygies. If $J, \widetilde{J} \in \mathfrak{A}$ we say that $\widetilde{J}$ *splits over* J when $\widetilde{J} \cong J \oplus T$ for some $T \in \mathfrak{S}$; we will prove the following.

Theorem 1.3. *Let $\mathbf{S}^{(n)}$ and $\widetilde{\mathbf{S}}^{(n)}$ be n-stems over M having syzygies $(J_r)_{1 \le r \le n}$ $(\widetilde{J}_r)_{1 \le r \le n}$ respectively. If $\mathbf{S}^{(n)}$ is minimal then $\widetilde{J}_r$ splits over J_r for $1 \le r \le n-1$.*

2. Some Categorical Preliminaries

We assume familiarity with the notions of *category* and *functor* [5]. We denote by $\mathcal{A}b$ the category of abelian groups and homomorphisms. In what follows we shall work with subcategories $\mathfrak{A}$ of $\mathcal{A}b$ which satisfy certain tameness conditions. These are defined formally below. However, it is instructive to consider them as they relate to two basic examples; thus suppose that Λ is a ring and consider

$\mathcal{M}od_\Lambda$: the category of right Λ-modules and Λ-homomorphisms:

By a graded Λ-module we mean a Λ-module M given as a direct sum $M = \oplus_{n \geq 0} M_n$ where each M_n is a Λ-submodule. A graded homomorphism $f : M \to N$ between two such graded modules is then a Λ-homomorphism satisfying $f(M_n) \subset N_n$ for each n and we may form

$\mathcal{G}(\Lambda)$: the category of *graded* right Λ-modules and Λ-homomorphisms.

Observe that $\mathcal{M}od_\Lambda$ may be regarded as the subcategory of $\mathcal{G}(\Lambda)$ consisting of graded modules in which $M_r = 0$ for $r > 0$. In turn, $\mathcal{G}(\Lambda)$ may be regarded as a subcategory of $\mathcal{A}b$ by forgetting both the grading and the Λ-structure. In the above examples, the following notions are well defined:

(i) Zero;
(ii) Kernels;
(iii) Images;
(iv) Exact sequences;
(v) Quotients.

In either case the nature of 'zero' should be obvious. Any module has a zero and hence a zero submodule. When $f : M \to N$ is a Λ-homomorphism

$$\operatorname{Ker}(f) = \{\mathbf{x} \in M \mid f(\mathbf{x}) = 0\}; \quad \operatorname{Im}(f) = \{f(\mathbf{x}) \mid \mathbf{x} \in M\}.$$

Then $\operatorname{Ker}(f)$ is a submodule of M and $\operatorname{Im}(f)$ a submodule of N. Moreover if f is a graded homomorphism then $\operatorname{Ker}(f)$ and $\operatorname{Im}(f)$ are graded by

$$\operatorname{Ker}(f)_n = \operatorname{Ker}(f) \cap M_n; \quad \operatorname{Im}(f)_n = \operatorname{Im}(f) \cap N_n.$$

A sequence of morphisms $A_1 \stackrel{\alpha_1}{\to} A_2 \stackrel{\alpha_2}{\to} \cdots \stackrel{\alpha_{n-1}}{\to} A_n \stackrel{\alpha_n}{\to} A_{n+1}$ is said to be *exact* when $\operatorname{Ker}(\alpha_{r+1}) = \operatorname{Im}(\alpha_r)$ for $1 \leq r \leq n - 1$. If $K \subset M$ is a Λ-submodule the quotient group M/K admits a natural

Λ-module structure. Moreover, if K is a graded submodule of the graded module M then M/K is graded by $(M/K)_n = M_n/K_n$. One may also construct

(vi) Pullbacks;
(vii) Direct products;
(viii) Pushouts;
(ix) Direct sums.

We first recall the definitions. If $\mathfrak{A}$ is a category and $f_i : M_i \to N$ are morphisms in $\mathfrak{A}$ ($i = 1, 2$) then by a *pullback* for f_1, f_2 we mean an object $\varprojlim(f_1, f_2)$ in $\mathfrak{A}$ together with morphisms $\pi_i : \varprojlim(f_1, f_2) \to M_i$ such that $f_1 \circ \pi_1 = f_2 \circ \pi_2$ which possess the following *universal property*:

If $\alpha_i : X \to M_i$ are morphisms in $\mathfrak{A}$ such that $f_1 \circ \alpha_1 = f_2 \circ \alpha_2$, then there exists a unique morphism $\alpha : X \to \varprojlim(f_1, f_2)$ making the following diagram commute

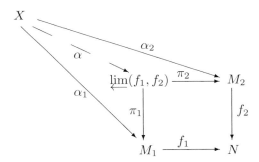

When $\varprojlim(f_1, f_2)$ exists the uniqueness condition on α guarantees that $\varprojlim(f_1, f_2)$ is unique up to isomorphism in $\mathfrak{A}$. We say that $\mathfrak{A}$ *has pullbacks* when $\varprojlim(f_1, f_2)$ exists for any pair of morphisms $f_i : M_i \to N$ ($i = 1, 2$). In $\mathcal{M}od_\Lambda$ pullbacks are defined by

$$\varprojlim(f_1, f_2) = \{(m_1, m_2) \in M_1 \times M_2 \mid f_1(m_1) = f_2(m_2)\},$$

where $\pi_i : \varprojlim(f_1, f_2) \to M_i$ is the obvious projection map. Moreover, in the special case where $N = 0$ the pullback construction simply yields the *direct product* $M_1 \times M_2$ showing that any pullback $\varprojlim(f_1, f_2)$ is a submodule of $M_1 \times M_2$. Note that in $\mathcal{G}(\Lambda)$ a direct product $M \times M'$ of graded modules admits a grading given by $(M \times M')_r = M_r \times M'_r$ which in turn induces a grading on any pullback contained therein.

Pushout is the dual notion to pullback. Here it is useful to recall that if $\mathfrak{A}$ is a category the *dual category* $\mathfrak{A}^*$ has the same objects and morphisms as $\mathfrak{A}$ but with the direction of all arrows reversed. One says that $\mathfrak{A}$ has *pushouts* when the dual category $\mathfrak{A}^*$ has pullbacks. Thus if $f_r : N \to M_i$ $(r = 1, 2)$ are morphisms in $\mathfrak{A}$ by a *pushout* for f_1, f_2 we mean an object $\varinjlim(f_1, f_2)$ in $\mathfrak{A}$ together with morphisms $i_r : M_r \to \varinjlim(f_1, f_2)$ such that $i_1 \circ f_1 = i_2 \circ f_2$ which possess the universal property which is dual to pullback. When $\varinjlim(f_1, f_2)$ exists the uniqueness condition on α again guarantees that $\varinjlim(f_1, f_2)$ is unique up to isomorphism. In the special case where $N = 0$ the pushout construction yields the *direct sum* $M_1 \oplus M_2$. In both $\mathcal{M}od_\Lambda$ and $\mathcal{G}(\Lambda)$ the direct sum $M_1 \oplus M_2$ coincides with the direct product $M_1 \times M_2$ with the canonical injections $i_r : M_r \to M_1 \oplus M_2$

$$i_1(\mathbf{x}) = (\mathbf{x}, 0); \quad i_2(\mathbf{x}) = (0, \mathbf{x}).$$

In $\mathcal{M}od_\Lambda$ $\varinjlim(f_1, f_2) = (M_1 \oplus M_2)/\mathrm{Im}(f_1 \oplus -f_2)$. Note that this module has a natural grading when f_1, f_2 are graded homomorphisms so that $\mathcal{G}(\Lambda)$ also has pushouts.

In what follows we work with categories $\mathfrak{A}$ in which the above notions **(i)**–**(ix)** are all present. Recall that a category $\mathfrak{A}$ is said to be *abelian* (cf. [4, 5]) when the following properties **(I)**, **(II)**, **(III)** hold[†]:

(I) $\mathfrak{A}$ has a zero object;

(II) $\mathfrak{A}$ has pullbacks and every monomorphism is a kernel;

(III) $\mathfrak{A}$ has pushouts and every epimorphism is a cokernel.

In any abelian category $\mathfrak{A}$ we define an addition on all $\mathrm{Hom}_\mathfrak{A}(A, B)$ thus:

$$+ : \mathrm{Hom}_\mathfrak{A}(A, B) \times \mathrm{Hom}_\mathfrak{A}(A, B) \to \mathrm{Hom}_\mathfrak{A}(A, B); \quad f + g = (f, g) \circ \delta,$$

where $\delta : A \to A \oplus A$ is the diagonal and the morphism $(f, g) : A \oplus A \to B$
is induced from $f : A \to B$ and $g : A \to B$ by regarding $A \oplus A$ as a pushout. When $\mathfrak{A}$ is an abelian category, we have the following additivity property whose proof is left as an exercise:

(x) $\mathrm{Hom}_\mathfrak{A}(A, B)$ is naturally an abelian group for any $A, B \in \mathfrak{A}$.

[†]We note (cf. [5, Chapter 1]) that there are many apparently different, though equivalent, ways of defining the notion of *abelian category*.

Recall that a category $\mathfrak{A}$ is said to be *small* when its objects form a set rather than merely a class. In this context, we note the following theorem of Lubkin (see [4, 5]).

Theorem 2.1. *If $\mathfrak{A}$ is a small abelian category there is a functor $\iota : \mathfrak{A} \to \mathcal{A}\mathrm{b}$ which preserves addition, exact sequences and for which $\mathrm{Hom}_{\mathfrak{A}}(A, B) \overset{i_*}{\rightarrowtail} \mathrm{Hom}_{\mathcal{A}\mathrm{b}}(\iota(A), \iota(B))$ is injective for all $A, B \in \mathfrak{A}$.*

Lubkin's theorem has the practical consequence that diagrams in any abelian category can be regarded simply as diagrams of additive abelian groups and homomorphisms; we take advantage of this in what follows.

By a *tame category* we mean one which is equivalent to a small subcategory of $\mathcal{A}\mathrm{b}$ and which is *abelian* in the above sense. In consequence we see that every small abelian category is tame. Evidently $\mathcal{M}\mathrm{od}_\Lambda$ and $\mathcal{G}(\Lambda)$ are abelian categories. However, without some size restriction on the objects neither category is tame. One especially convenient restriction is to consider only rings Λ which are countable. We then denote by $\mathcal{M}\mathrm{od}_\Lambda^\infty$ the full subcategory of $\mathcal{M}\mathrm{od}_\Lambda$ consisting of *countably generated* modules. Likewise $\mathcal{G}^\infty(\Lambda)$ will denote the full subcategory of $\mathcal{G}(\Lambda)$ whose underlying modules are countably generated. It follows easily from the following proposition.

Proposition 2.2. *If Λ is a countable ring, then $\mathcal{M}\mathrm{od}_\Lambda^\infty$ and $\mathcal{G}^\infty(\Lambda)$ are tame abelian categories.*

3. Splitting and Projectives

In what follows, $\mathfrak{A}$ will denote a tame abelian category. We recall the following basic result, the *Five Lemma* which, via Lubkin's theorem, it suffices to prove in $\mathcal{A}\mathrm{b}$.

Suppose given a commutative diagram in $\mathfrak{A}$ with exact rows

$$\begin{array}{ccccccccc} A_1 & \overset{\alpha_1}{\to} & A_2 & \overset{\alpha_2}{\to} & A_3 & \overset{\alpha_3}{\to} & A_4 & \overset{\alpha_4}{\to} & A_5 \\ f_1 \downarrow & & f_2 \downarrow & & f_3 \downarrow & & f_4 \downarrow & & f_5 \downarrow \\ B_1 & \overset{\beta_1}{\to} & B_2 & \overset{\beta_2}{\to} & B_3 & \overset{\beta_3}{\to} & B_4 & \overset{\beta_4}{\to} & B_5 \end{array} \qquad (3.1)$$

If f_1, f_2, f_4 and f_5 are all isomorphisms, then f_3 is also an isomorphism.

Given objects $A, C \in \mathfrak{A}$, there are canonical morphisms $i_A : A \to A \oplus C$ and $\pi_C : A \oplus C \to C$ allowing the construction of the *trivial exact sequence*

$$\mathcal{T} = (0 \to A \xrightarrow{i_A} A \oplus C \xrightarrow{\pi_C} C \to 0).$$

An exact sequence $\mathcal{E} = (0 \to C \xrightarrow{i} B \xrightarrow{p} A \to 0)$ in $\mathfrak{A}$ is said to *split* when it is isomorphic to the trivial exact sequence by means of a commutative diagram as follows:

$$\begin{array}{ccccccccc} 0 \to & A & \xrightarrow{i} & B & \xrightarrow{p} & C & \to 0 \\ & \downarrow \mathrm{Id}_A & & \downarrow \psi & & \downarrow \mathrm{Id}_C & \\ 0 \to & A & \xrightarrow{i_A} & A \oplus C & \xrightarrow{\pi_C} & C & \to 0. \end{array}$$

It follows from the Five Lemma that such a splitting ψ is necessarily an isomorphism. We say that $\mathcal{E}$ *splits on the left* when there exists a morphism $r : B \to A$ such that $r \circ i = \mathrm{Id}_A$. Finally we say that $\mathcal{E}$ *splits on the right* when there exists a morphism $s : A \to B$ such that $p \circ s = \mathrm{Id}_C$. If ψ is a splitting of $\mathcal{E}$ then $r = \pi_A \circ \psi$ is a left splitting of $\mathcal{E}$. Conversely if $r : B \to C$ is a left splitting of $\mathcal{E}$ then $\psi = \binom{r}{p} : B \to A \oplus C$ is a splitting. If ψ is a splitting of $\mathcal{E}$ then $s = \psi^{-1} \circ i_C : C \to B$ is a right splitting. Conversely, if s is a right splitting then by the Five Lemma, $(i, s) : A \oplus C \to B$ is necessarily an isomorphism and $\psi = (i, s)^{-1}$ is then a splitting. To summarise:

$$\mathcal{E} \text{ splits} \iff \mathcal{E} \text{ splits on the left} \iff \mathcal{E} \text{ splits on the right}. \qquad (3.2)$$

We say that an object $Q \in \mathfrak{A}$ is *projective* when every exact sequence of the form $0 \to C \xrightarrow{i} B \xrightarrow{p} Q \to 0$ splits. The following proposition is fundamental:

Proposition 3.1 (Schanuel's Lemma). *Let* $(0 \to D_r \xrightarrow{i_r} P_r \xrightarrow{f_r} M \to 0)$ *be exact sequences in* Mod_Λ *($r = 1, 2$). If P_1 and P_2 are projective, then $D_1 \oplus P_2 \cong D_2 \oplus P_1$.*

Proof. Form the pullback $Q = \varprojlim(f_1, f_2)$ Then there is a short exact sequence $0 \to D_2 \to Q \xrightarrow{\pi_1} P_1 \to 0$ that splits as P_1 is projective. Hence $Q \cong D_2 \oplus P_1$. Likewise the short exact sequence $0 \to D_1 \to Q \xrightarrow{\pi_2} P_2 \to 0$ splits as P_2 is projective. Thus $D_1 \oplus P_2 \cong Q \cong D_2 \oplus P_1$ as claimed. $\square$

4. Some Standard Diagrams

Consider the following commutative diagram in a tame abelian category $\mathfrak{A}$ in which it is assumed that all rows and columns are exact:

$$\begin{cases} & & & & 0 & \\ & & & & \downarrow & \\ & S_- & \xrightarrow{\widehat{i}} & S & \xrightarrow{\widehat{p}} & S_+ & \\ & \downarrow j_- & & \downarrow j & & \downarrow j_+ & \\ & \widetilde{K} & \xrightarrow{\widetilde{i}} & \widetilde{F} & \xrightarrow{\widetilde{p}} & \widetilde{J} & \\ & \downarrow \varphi_- & & \downarrow \varphi & & \downarrow \varphi_+ & \\ 0 \to & K & \xrightarrow{i} & F & \xrightarrow{p} & J & \to 0 \\ & & & & \downarrow & \\ & & & & 0 & \end{cases} \quad (4.1)$$

By Lubkin's Theorem we may replace it by an equivalent diagram in $\mathcal{A}b$. A straightforward diagram chase then shows that, in (4.1):

φ and $\widehat{p}$ are both epimorphic $\iff$ $\widetilde{p}$ and φ_- are both epimorphic. (4.2)

Consider likewise

$$\begin{cases} & & 0 & & 0 & & 0 & \\ & & \downarrow & & \downarrow & & \downarrow & \\ & & C_2 & \xrightarrow{\gamma_2} & C_1 & \xrightarrow{\gamma_1} & C_0 & \\ & & \downarrow j_2 & & \downarrow j_1 & & \downarrow j_0 & \\ B_3 & \xrightarrow{\beta_3} & B_2 & \xrightarrow{\beta_2} & B_1 & \xrightarrow{\beta_1} & B_0 & \\ \downarrow \varphi_3 & & \downarrow \varphi_2 & & \downarrow \varphi_1 & & \downarrow \varphi_0 & \\ A_3 & \xrightarrow{\alpha_3} & A_2 & \xrightarrow{\alpha_2} & A_1 & \xrightarrow{\alpha_1} & A_0 & \\ \downarrow & & \downarrow & & \downarrow & & \downarrow & \\ 0 & & 0 & & 0 & & 0 & \end{cases} \quad (4.3)$$

Suppose in (4.3) that the columns are all exact. If the rows

$$(A_3 \xrightarrow{\alpha_3} A_2 \xrightarrow{\alpha_2} A_1 \xrightarrow{\alpha_1} A_0) \quad \text{and} \quad (B_3 \xrightarrow{\beta_3} B_2 \xrightarrow{\beta_2} B_1 \xrightarrow{\beta_1} B_0) \quad (4.4)$$

are exact then $(C_2 \xrightarrow{\gamma_2} C_1 \xrightarrow{\gamma_1} C_0)$ is also exact.

In the following commutative diagram $\mathcal{C}$ over $\mathfrak{A}$ we assume all rows and columns are exact:

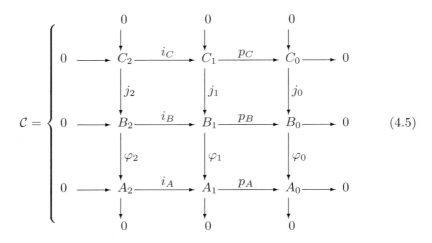 (4.5)

We say that the diagram $\mathcal{C}$ of (4.5) *splits completely* when there are morphisms $r_t : B_t \to C_t$ for $t = 0, 1, 2$ such that $r_t \circ j_t = \mathrm{Id}_{C_t}$ and such that the following diagram commutes

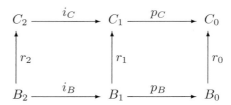

The triple (r_0, r_1, r_2) is then called a *complete splitting of* $\mathcal{C}$. Evidently r_0 is a (left) splitting of the exact sequence

$$0 \to C_0 \xrightarrow{j_0} B_0 \xrightarrow{\varphi_0} A_0 \to 0$$

and we say the complete splitting (r_0, r_1, r_2) *extends* the splitting r_0.

Theorem 4.1. *Assume in (4.5) above that all rows and columns are exact and that A_1 and C_0 are projective. Then any (left) splitting r_0 of the right-hand column extends to a complete splitting (r_0, r_1, r_2) of $\mathcal{C}$.*

5. A Comparison Theorem for Resolutions

Given an integer $n \geq 0$ and an object $M \in \mathfrak{A}$, by an *n-resolution* we mean an exact sequence in $\mathfrak{A}$ of the form

$$\mathbf{E}^{(n)} = (E_n \xrightarrow{\partial_n} E_{n-1} \xrightarrow{\partial_{n-1}} \cdots \xrightarrow{\partial_2} E_1 \xrightarrow{\partial_1} E_0 \xrightarrow{\epsilon} M \to 0).$$

We allow ourselves to write $E_{-1} = M$ and $\partial_0 = \epsilon$ whenever it is notationally convenient to do so. Whilst later we shall require the resolving objects E_r to be projective of a special type, here we impose no restriction other than exactness. We denote by $\mathfrak{A}(n)$ the category whose objects are such sequences and in which morphisms are commutative ladders

$$\begin{array}{c} \widetilde{\mathbf{E}}^{(n)} \\ \varphi \downarrow \\ \mathbf{E}^{(n)} \end{array} = \left(\begin{array}{ccccccccc} \widetilde{E}_n & \xrightarrow{\widetilde{\partial}_n} & \widetilde{E}_{n-1} & \xrightarrow{\widetilde{\partial}_{n-1}} & \cdots & \xrightarrow{\widetilde{\partial}_1} & \widetilde{E}_0 & \xrightarrow{\widetilde{\epsilon}} & \widetilde{M} & \to 0 \\ \varphi_n \downarrow & & \varphi_{n-1} \downarrow & & & & \varphi_0 \downarrow & & \downarrow \varphi_{-} & \\ E_n & \xrightarrow{\partial_n} & E_{n-1} & \xrightarrow{\partial_{n-1}} & \cdots & \xrightarrow{\partial_1} & E_0 & \xrightarrow{\epsilon} & M & \to 0 \end{array} \right).$$

We also allow the limiting case $n = \infty$ in the obvious way. If $\varphi_{-} : \widetilde{M} \to M$ is an epimorphism we say that φ is a *dominating morphism over* φ_{-} when each φ_r is also an epimorphism. We agree to write $\mathbf{E}^{(n)} \preceq \widetilde{\mathbf{E}}^{(n)}$ in the special case where $\varphi_{-} = \mathrm{Id}_M : M \to M$.

For the rest of this section we pick a specific dominating morphism $\varphi : \widetilde{\mathbf{E}}^{(n)} \to \mathbf{E}^{(n)}$ over Id_M. Defining $T_r = \mathrm{Ker}(\varphi_r)$, $j_r : T_r \to \widetilde{E}_r$ will denote the 'inclusion' and $\widehat{\partial}_r : T_r \to T_{r-1}$ the 'restriction' giving a commutative diagram:

$$\left\{ \begin{array}{ccccccccc} T_n & \xrightarrow{\widehat{\partial}_n} & T_{n-1} & \xrightarrow{\widehat{\partial}_{n-1}} & \cdots & \xrightarrow{\widehat{\partial}_1} & T_0 & \to 0 & \to 0 \\ j_n \downarrow & & j_{n-1} \downarrow & & & & j_0 \downarrow & & \downarrow \\ \widetilde{E}_n & \xrightarrow{\widetilde{\partial}_n} & \widetilde{E}_{n-1} & \xrightarrow{\widetilde{\partial}_{n-1}} & \cdots & \xrightarrow{\widetilde{\partial}_1} & \widetilde{E}_0 & \xrightarrow{\widetilde{\epsilon}} & M \to 0 \\ \varphi_n \downarrow & & \varphi_{n-1} \downarrow & & & & \varphi_0 \downarrow & & \downarrow \\ E_n & \xrightarrow{\partial_n} & E_{n-1} & \xrightarrow{\partial_{n-1}} & \cdots & \xrightarrow{\partial_1} & E_0 & \xrightarrow{\epsilon} & M \to 0. \end{array} \right. \quad (5.1)$$

Although $\widehat{\partial}_{n-1} \circ \widehat{\partial}_n = 0$ it is not, in general, true that $\mathrm{Ker}(\widehat{\partial}_{n-1})$ is the same as $\mathrm{Im}(\widehat{\partial}_n)$. Noting this loss of information at the top left-hand corner, it nevertheless follows, by induction from (4.4), that the following portion of (5.1) has exact rows and columns:

$$
\left\{
\begin{array}{ccccccccccc}
& & 0 & & 0 & & & & 0 & & 0 \\
& & \downarrow & & \downarrow & & & & \downarrow & & \downarrow \\
& & T_{n-1} & \overset{\widehat{\partial}_{n-1}}{\to} & T_{n-2} & \overset{\widehat{\partial}_{n-2}}{\to} & \cdots & \overset{\widehat{\partial}_1}{\to} & T_0 & \to 0 & \to 0 \\
& & j_{n-1} \downarrow & & j_{n-2} \downarrow & & & & j_0 \downarrow & & \downarrow \\
\widetilde{E}_n & \overset{\widetilde{\partial}_n}{\to} & \widetilde{E}_{n-1} & \overset{\widetilde{\partial}_{n-1}}{\to} & \widetilde{E}_{n-2} & \overset{\widetilde{\partial}_{n-2}}{\to} & \cdots & \overset{\widetilde{\partial}_1}{\to} & \widetilde{E}_0 & \overset{\widetilde{\epsilon}}{\to} M & \to 0 \\
\varphi_n \downarrow & & \varphi_{n-1} \downarrow & & \varphi_{n-2} \downarrow & & & & \varphi_0 \downarrow & & \downarrow \\
E_n & \overset{\partial_n}{\to} & E_{n-1} & \overset{\partial_{n-1}}{\to} & E_{n-2} & \overset{\partial_{n-2}}{\to} & \cdots & \overset{\partial_1}{\to} & E_0 & \overset{\epsilon}{\to} M & \to 0 \\
\downarrow & & \downarrow & & \downarrow & & & & \downarrow & & \downarrow \\
0 & & 0 & & 0 & & & & 0 & & 0
\end{array}
\right.
$$
(5.2)

In the above we define $J_r = \mathrm{Ker}(\partial_{r-1})$ for $1 \leq r \leq n+1$. When $r \leq n$ then it is also true that $J_r = \mathrm{Im}(\partial_r)$ and we then write $\partial_r = i_r \circ p_r$ for the canonical decomposition of ∂_r through its image with i_r monomorphic and p_r epimorphic:

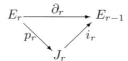

Likewise we consider the corresponding decompositions for the $\widetilde{\partial}_r$ to obtain commutative diagrams as follows:

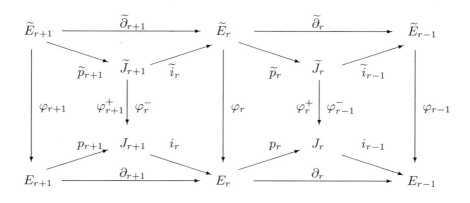

(5.3)

where, depending on context, both φ_r^- and φ_{r-1}^+ denote the restriction $\varphi_{r-1|\widetilde{J}_r} : \widetilde{J}_r \to J_r$. Now taking the corresponding decompositions for the $\widehat{\partial}_r$ we get commutative diagrams

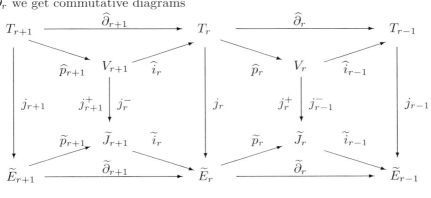

(5.4)

where j_r^- and j_{r-1}^+ denote the 'inclusion' $V_r \to \widetilde{J}_r$. We assemble (5.3) and (5.4) into commutative diagrams $\mathcal{D}(r)$ for $1 \leq r \leq n-2$;

$$\mathcal{D}(r) = \begin{cases} \begin{array}{ccccccc} & 0 & & 0 & & 0 & \\ & \downarrow & & \downarrow & & \downarrow & \\ 0 \to & V_{r+1} & \stackrel{\widehat{i_r}}{\to} & T_r & \stackrel{\widehat{p_r}}{\to} & V_r & \to 0 \\ & \downarrow j_r^+ & & \downarrow j_r & & \downarrow j_r^- & \\ 0 \longrightarrow & \widetilde{J_{r+1}} & \stackrel{\widetilde{i_r}}{\to} & \widetilde{E}_r & \stackrel{\widetilde{p_r}}{\to} & \widetilde{J}_r & \longrightarrow 0 \\ & \downarrow \varphi_r^+ & & \downarrow \varphi_r & & \downarrow \varphi_r^- & \\ 0 \longrightarrow & J_{r+1} & \stackrel{i_r}{\to} & E_r & \stackrel{p_r}{\longrightarrow} & J_r & \longrightarrow 0 \\ & \downarrow & & \downarrow & & \downarrow & \\ & 0 & & 0 & & 0 & \end{array} \end{cases}$$

In the special case $r = 0$ we obtain

$$\mathcal{D}(0) = \begin{cases} \begin{array}{ccccccc} & 0 & & 0 & & & \\ & \downarrow & & \downarrow & & & \\ 0 \to & V_1 & = & T_0 & \to & 0 & \\ & \downarrow j_0^+ & & \downarrow j_0 & & \downarrow & \\ 0 \to & \widetilde{J}_1 & \stackrel{\widetilde{i_0}}{\longrightarrow} & \widetilde{E}_0 & \stackrel{\widetilde{\epsilon}}{\longrightarrow} & M & \to 0 \\ & \downarrow \varphi_0^+ & & \downarrow \varphi_0 & & \| \text{ Id} & \\ 0 \to & J_1 & \stackrel{i_0}{\longrightarrow} & E_0 & \stackrel{\epsilon}{\longrightarrow} & M & \to 0 \\ & \downarrow & & \downarrow & & \downarrow & \\ & 0 & & 0 & & 0 & \end{array} \end{cases}$$

As $\widetilde{\epsilon} \circ j_0 = \epsilon \circ \varphi_0 \circ j_0 = 0$ then the 'inclusion' $j_0^+ : V_1 \to \widetilde{J}_1 = \mathrm{Ker}(\widetilde{\epsilon})$ and 'restriction' $\varphi_0^+ : \varphi_{1|\widetilde{J}_1} : \widetilde{J}_1 \to J_1$ are both well defined.

Proposition 5.1. *All the rows and columns of $\mathcal{D}(0)$ are exact.*

Proof. Exactness of the rows and of the right-hand and middle columns is tautological. Thus it suffices to show that:

(a) φ_0^+ is epimorphic and (b) $\mathrm{Ker}(\varphi_0^+) = \mathrm{Im}(j_0^+)$.

For (a), observe that in the following subdiagram of $\mathcal{D}(0)$ all rows and columns are exact:

$$\begin{array}{ccccccc}
 & & & & T_0 & \xrightarrow{\widehat{p_0}} & 0 \\
 & & & & \downarrow j_0 & & \downarrow \\
 & & \widetilde{J}_1 & \xrightarrow{\widetilde{i_0}} & \widetilde{E}_0 & \xrightarrow{\widetilde{\epsilon}} & M \\
 & & \downarrow \varphi_0^+ & & \downarrow \varphi_0 & & \| \mathrm{Id} \\
0 & \longrightarrow & J_1 & \xrightarrow{i_0} & E_0 & \xrightarrow{\epsilon} & M & \to 0 \\
 & & & & \downarrow & & \downarrow \\
 & & & & 0 & & 0
\end{array}$$

As both φ_0 and $\widehat{p}_0$ are epimorphic then φ_0^+ is epimorphic by (4.2).

To prove (b) we may again, by Lubkin's theorem, assume that the diagram consists of abelian groups and homomorphisms in which monomorphisms become inclusions thus:

$$\begin{array}{ccccccc}
 & & 0 & & 0 & & \\
 & & \downarrow & & \downarrow & & \\
0 \to & & V_1 & = & T_0 & \to & 0 \\
 & & \cap j_0^+ & & \cap j_0 & & \downarrow \\
0 \to & & \widetilde{J}_1 & \xrightarrow{\widetilde{i_0}} & \widetilde{E}_0 & \xrightarrow{\widetilde{\epsilon}} & M \to 0 \\
 & & \downarrow \varphi_0^+ & & \downarrow \varphi_0 & & \| \mathrm{Id} \\
0 \to & & J_1 & \xrightarrow{i_0} & E_0 & \xrightarrow{\epsilon} & M \to 0 \\
 & & \downarrow & & \downarrow & & \downarrow \\
 & & 0 & & 0 & & 0
\end{array}$$

The inclusion $\mathrm{Im}(j_0^+) \subset \mathrm{Ker}(\varphi_0^+)$ then follows by restriction from $\varphi_0 \circ j_0 = 0$. Thus suppose $x \in \widetilde{J}_1$ satisfies $\varphi_0^+(x) = 0$; then $\widetilde{i}_0(x) \in \mathrm{Ker}(\varphi_0) = T_0 = V_1$, completing the proof. $\square$

Before proceeding we first note:

the rows of each $\mathcal{D}(r)$ are exact; (5.5)

the middle column of each $\mathcal{D}(r)$ is exact; (5.6)

the right-hand column of $\mathcal{D}(r)$ is identical
to the left-hand column of $\mathcal{D}(r-1)$ for $1 \leq r \leq n-1$. (5.7)

We arrive at the following 'weak comparison' theorem.

Theorem 5.2. *Let $\varphi : \widetilde{\mathbf{E}}^{(n)} \to \mathbf{E}^{(n)}$ be a dominating morphism over Id_M where $n \geq 2$. Then the rows and columns of $\mathcal{D}(r)$ are exact for $0 \leq r \leq n-2$.*

Proof. For $n = 2$ this is simply a restatement of Theorem 5.1. Thus we may suppose that $n \geq 3$. Let $\mathbf{C}(r)$ be the statement that the rows and columns of $\mathcal{D}(r)$ are exact. As $\mathbf{C}(0)$ is true, again by Theorem 5.1, it suffices to show that $\mathbf{C}(r-1) \Rightarrow \mathbf{C}(r)$ for $1 \leq r \leq n-2$.

Via the Lubkin imbedding it suffices to prove the statement for the corresponding diagram of abelian groups and homomorphisms. By induction the right-hand column of $\mathcal{D}(r)$ is exact as it coincides with the left-hand column of $\mathcal{D}(r-1)$. As observed in (5.6) the middle column of $\mathcal{D}(r)$ is exact so it suffices to show that the left-hand column of $\mathcal{D}(r)$ is exact. As j_r^+ is a monomorphism it suffices to show:

(a) φ_r^+ is epimorphic;
(b) $\mathrm{Ker}(\varphi_r^+) = \mathrm{Im}(j_r^+)$.

To show (a), note that in the following subdiagram of $\mathcal{D}(r)$ all rows and columns are exact:

$$\begin{array}{ccccc}
 & & 0 & & \\
 & & \downarrow & & \\
 & & T_r & \xrightarrow{\widehat{p_r}} & V_r \longrightarrow 0 \\
 & & \downarrow j_r & & \downarrow j_r^- \\
\widetilde{J}_{r+1} & \xrightarrow{\widetilde{i_r}} & \widetilde{E}_r & \xrightarrow{\widetilde{p_r}} & \widetilde{J}_r \longrightarrow 0 \\
\downarrow \varphi_r^+ & & \downarrow \varphi_r & & \downarrow \varphi_r^- \\
J_{r+1} & \xrightarrow{i_r} & E_r & \xrightarrow{p_r} & J_r \longrightarrow 0 \\
 & & \downarrow & & \downarrow \\
 & & 0 & & 0
\end{array}$$

As φ_r and $\widehat{p}_r$ are epimorphic it follows by (4.2) that φ_r^+ is epimorphic.

To prove (b) suppose $x \in \widetilde{J}_{r+1} = \mathrm{Ker}(\widehat{\partial}_r)$ satisfies $\varphi_r^+(x) = 0$. We must produce an element $y \in V_{r+1} = \mathrm{Ker}(\widehat{\partial}_r)$ such that $j_r(y) = x$. Consider the following portion of the diagram established in (5.2). Observe that as $r \leq n - 2$ this subdiagram is well defined.

$$\begin{array}{ccccccc}
& & T_{r+1} & \stackrel{\widehat{\partial}_{r+1}}{\longrightarrow} & T_r & \stackrel{\widehat{\partial}_r}{\longrightarrow} & T_{r-1} \\
& & j_{r+1} \downarrow & & j_r \downarrow & & j_{r-1} \downarrow \\
\widetilde{E}_{r+2} & \stackrel{\widetilde{\partial}_{r+2}}{\longrightarrow} & \widetilde{E}_{r+1} & \stackrel{\widetilde{\partial}_{r+1}}{\longrightarrow} & \widetilde{E}_r & \stackrel{\widetilde{\partial}_r}{\longrightarrow} & \widetilde{E}_{r-1} \\
\varphi_{r+2} \downarrow & & \varphi_{r+1} \downarrow & & \varphi_r \downarrow & & \varphi_{r-1} \downarrow \\
E_{r+2} & \stackrel{\partial_{r+2}}{\longrightarrow} & E_{r+1} & \stackrel{\partial_{r+1}}{\longrightarrow} & E_r & \stackrel{\partial_r}{\longrightarrow} & E_{r-1} \\
\downarrow & & & & & & \\
0 & & & & & &
\end{array}$$

The conditions on $x \in \widetilde{E}_r$ are $\widetilde{\partial}_r(x) = 0$ and $\varphi_r(x) = 0$. By exactness of the middle row we may choose $w \in \widetilde{E}_{r+1}$ such that $\widetilde{\partial}_{r+1}(w) = x$. Then $\varphi_r \circ \widetilde{\partial}_{r+1}(w) = 0$ so that $\partial_{r+1} \circ \varphi_{r+1}(w) = 0$. By exactness of the bottom row choose $z \in E_{r+2}$ such that $\partial_{r+2}(z) = \varphi_{r+1}(w)$.

As $\varphi_{r+2} : \widetilde{E}_{r+2} \to E_{r+2}$ is epimorphic, choose $\zeta \in \widetilde{E}_{r+2}$ such that $\varphi_{r+2}(\zeta) = z$; then $\partial_{r+2} \circ \varphi_{r+2}(\zeta) = \varphi_{r+1}(w)$. Put $\mu = w - \widetilde{\partial}_{r+2}(\zeta) \in \widetilde{E}_{r+1}$ so that $\varphi_{r+1}(\mu) = 0$. Choose $\eta \in T_{r+1}$ such that

$$j_{r+1}(\eta) = \mu = w - \widetilde{\partial}_{r+2}(\zeta).$$

Then $\widetilde{\partial}_{r+1} \circ j_{r+1}(\eta) = \widetilde{\partial}_{r+1}(w) - \widetilde{\partial}_{r+1}\widetilde{\partial}_{r+2}(\zeta)$. Put $y = \widehat{\partial}_{r+1}(\eta)$. Then $y \in V_{r+1}$ and $j_r(y) = x$. This completes the proof. $\square$

The statement of Theorem 5.2 extends to the limiting case $n = \infty$ as follows.

Corollary 5.3. *Let* $\varphi : \widetilde{\mathbf{E}}^{(\infty)} \to \mathbf{E}^{(\infty)}$ *be a dominating morphism over* Id_M. *Then the rows and columns of* $\mathcal{D}(r)$ *are exact for all* $r \geq 0$.

6. Finiteness Conditions and Stability

Let $\mathfrak{A}$ be a tame abelian category. By a *special class* in $\mathfrak{A}$, we mean a class of objects $\mathfrak{S} \subset \mathfrak{A}$ satisfying the following properties $\mathfrak{S}(1)$–$\mathfrak{S}(3)$:

$\mathfrak{S}(1)$: Each $S \in \mathfrak{S}$ is projective and $0 \in \mathfrak{S}$;

$\mathfrak{S}(2)$: If $0 \to X \to Y \to S \to 0$ is exact in $\mathfrak{A}$ and $S \in \mathfrak{S}$ then

$$X \in \mathfrak{S} \Leftrightarrow Y \in \mathfrak{S}.$$

Finally we have a 'finiteness' property. If $S, T \in \mathfrak{S}$ let $\mathbf{e}(S,T)$ denote the set of integers k for which there exists an epimorphism $S \to \underbrace{T \oplus \cdots \oplus T}_{k}$.

$\mathfrak{S}(3)$: If $S, T \in \mathfrak{S}$ and $T \neq 0$ then $\mathbf{e}(S,T)$ is bounded above.

It follows from $\mathfrak{S}(1)$ and $\mathfrak{S}(2)$ that $\mathfrak{S}$ is closed with respect to coproducts;

$$X \in \mathfrak{S} \quad \text{and} \quad Y \in \mathfrak{S} \Longrightarrow X \oplus Y \in \mathfrak{S}. \tag{6.1}$$

Likewise $\mathfrak{S}$ is closed with respect to isomorphism:

$$X \in \mathfrak{S} \quad \text{and} \quad X \cong_{\mathfrak{A}} Y \Longrightarrow Y \in \mathfrak{S}. \tag{6.2}$$

Recall that a finitely generated module M over a ring Λ is said to be *stably free* when $M \oplus \Lambda^m \cong \Lambda^{m+n}$ for some integers m, n.

The class $\mathcal{SF}$ of finitely generated stably free Λ-modules is a special class in $\mathcal{M}od_\Lambda$. (6.3)

Similarly we define a class $\mathcal{GSF}$ of objects in $\mathcal{G}(\Lambda)$ as follows:

$$M \in \mathcal{GSF} \iff \text{each } M_r \text{ is finitely generated stably free over } \Lambda. \tag{6.4}$$

The class $\mathcal{GSF}$ is a special class in $\mathcal{G}(\Lambda)$.

There is a relation, $\mathfrak{S}$-equivalence, defined on the objects of $\mathfrak{A}$ by

$$X \sim X' \iff X \oplus S \cong X' \oplus S' \quad \text{for some } S, S' \in \mathfrak{S}.$$

We define a class $\mathcal{F}(0)$ of objects in $\mathfrak{A}$ as follows: $M \in \mathcal{F}(0)$ when there exists an epimorphism $\eta : S \to M$ for some $S \in \mathfrak{S}$.

Proposition 6.1. *Let $M \in \mathcal{F}(0)$. If $T \in \mathfrak{S}$ is such that $M \oplus T \cong M$ then $T = 0$.*

Proof. Let $\varphi : S \to M$ be an epimorphism where $S \in \mathfrak{S}$, and suppose that there is an isomorphism $\psi_1 : M \to M \oplus T$ where $T \in \mathfrak{S}$. Then for each positive integer k we obtain an isomorphism $\psi_k : M \to M \oplus T^{(k)}$ on putting $\psi_k = (\psi_{k-1} \oplus \mathrm{Id}_T) \circ \psi_1$ for $k \geq 2$. Hence for each positive integer k we obtain an epimorphism $\eta_k : S \to T^{(k)}$ on putting $\eta_k = \pi_k \circ \psi_k \circ \varphi$. This contradicts property $\mathfrak{S}(3)$ unless $T = 0$. $\square$

Corollary 6.2. *Let $S \in \mathfrak{S}$. If $\varphi : S \to S$ is an epimorphism then φ is an isomorphism.*

Proof. Suppose that $\varphi : S \to S$ is an epimorphism. As S is projective then from the exact sequence

$$0 \to \mathrm{Ker}(\varphi) \to S \xrightarrow{\varphi} S \to 0$$

there is an isomorphism $\psi_1 : S \to S \oplus \mathrm{Ker}(\varphi)$ and $\mathrm{Ker}(\varphi) \in \mathfrak{S}$. By Proposition 6.1 $\mathrm{Ker}(\varphi) = 0$, so that φ is monomorphic and hence an isomorphism. $\square$

We first introduce a general definition. If $M_1, M_2 \in \mathcal{F}(0)$ we say that M_2 *splits over* M_1, written $M_1 \dashv M_2$, when there is an isomorphism $M_1 \oplus T \cong M_2$ in which $T \in \mathfrak{S}$. Evidently one has the following proposition.

Proposition 6.3. *If $M_1 \dashv M_2$ then $M_1 \sim M_2$.*

It is straightforward to see that the relation '$\dashv$' is transitive; that is:

$$\text{If } M_1 \dashv M_2 \text{ and } M_2 \dashv M_3 \text{ then } M_1 \dashv M_3. \tag{6.5}$$

More subtly, '$\dashv$' is also anti-symmetric in the sense that, for $M_1, M_2 \in \mathcal{F}(0)$, we have the following.

Proposition 6.4. $M_1 \dashv M_2 \wedge M_2 \dashv M_1 \Rightarrow M_1 \cong M_2.$

Proof. The hypothesis allows us to write $M_2 \cong M_1 \oplus T_1$ and $M_1 \cong M_2 \oplus T_2$ for some $T_1, T_2 \in \mathfrak{S}$. Thus $M_1 \cong M_1 \oplus T$ where $T = (T_1 \oplus T_2) \in \mathfrak{S}$. It follows from Corollary 6.2 above that $T = 0$. Hence $T_2 = 0$ and $M_1 \cong M_2$. $\square$

Corollary 6.5. *If Ω is an $\mathfrak{S}$-class of type $\mathcal{F}(0)$ then the relation '$\dashv$' induces a partial ordering on the isomorphism types of Ω.*

If X is an object in $\mathfrak{A}$, the $\mathcal{S}$-class $[X]$ is defined to be the collection of isomorphism classes of objects Y in $\mathfrak{A}$ which are $\mathfrak{S}$-equivalent to X:

$$[X] = \{Y \in \mathfrak{A} \mid Y \sim X\}/\cong.$$

As $\mathfrak{A}$ is equivalent to a small subcategory of $\mathcal{A}b$ it follows that

$$[X] \text{ is a set for each object } X \in \mathfrak{A}. \tag{6.6}$$

We denote by $\mathfrak{S}(n)$ the full subcategory of $\mathfrak{A}(n)$ consisting of exact sequences of the form

$$\mathbf{S}^{(n)} = (S_n \xrightarrow{\partial_n} S_{n-1} \xrightarrow{\partial_{n-1}} \cdots \xrightarrow{\partial_2} S_1 \xrightarrow{\partial_1} S_0 \xrightarrow{\epsilon} M \to 0)$$

in which $S_0, \ldots, S_n \in \mathfrak{S}$. Such a sequence will be called an *n-stem* over M. Moreover $J_{r+1} = \mathrm{Ker}(\partial_r)$ is called the $(r+1)$th *syzygy* of $\mathbf{S}^{(n)}$. We say that

M is of type $\mathcal{F}(n)$ when there exists an n-stem over M. In general this condition is a non-trivial restriction on M.

Theorem 6.6. *Let* $M \in \mathfrak{A}$ *and* $S \in \mathfrak{S}$. *Then*

$$M \in \mathcal{F}(n) \iff M \oplus S \in \mathcal{F}(n).$$

Proof. Let $\mathcal{P}(n)$ be the statement of the theorem; we first prove $\mathcal{P}(0)$. Suppose that $\epsilon : S_0 \to M$ is an epimorphism. Then $\epsilon \oplus \mathrm{Id} : S_0 \oplus S \to M \oplus S$ is also an epimorphism so that if $M \in \mathcal{F}(0)$ then $M \oplus S \in \mathcal{F}(0)$. Conversely suppose that $M \oplus S \in \mathcal{F}(0)$ and let $\eta : S_0 \to M \oplus S$ be an epimorphism. Taking $\pi_1 : M \oplus S \to M$, $\pi_2 : M \oplus S \to S$ to be the canonical projections, put $S' = \mathrm{Ker}(\pi_2 \circ \eta)$. Applying τ we obtain an exact sequence

$$0 \to \tau(S') \to \tau(S_0) \xrightarrow{\tau(\pi_1 \circ \eta)} \tau(S) \to 0.$$

It follows that the sequence $0 \to S' \to S_0 \xrightarrow{\pi_1 \circ \epsilon} S \to 0$ is also exact so that $S' \in \mathfrak{S}$ by property $\mathfrak{S}(3)$. However, $\tau(\pi_1 \circ \eta) : \tau(S') \to \tau(M)$ is epimorphic so that $\pi_1 \circ \eta : S' \to M$ is epimorphic and hence $M \in \mathcal{F}(0)$. This proves $\mathcal{P}(0)$. Now suppose that $\mathcal{P}(n-1)$ is true for $n \geq 1$, let $M \in \mathcal{F}(n)$ and let

$$S_n \xrightarrow{\partial_n} \cdots \xrightarrow{\partial_1} S_0 \xrightarrow{\epsilon} M \to 0$$

be an n-stem. Letting $i : S_0 \to S_0 \oplus S$ be the canonical morphism define

$$\delta_r = \begin{cases} i \circ \partial_1, & r = 1, \\ \partial_r, & r > 1. \end{cases}$$

We see that $S_n \xrightarrow{\delta_n} \cdots \xrightarrow{\delta_1} S_0 \oplus S \xrightarrow{\epsilon \oplus \mathrm{Id}} M \oplus S \to 0$ is exact. As $S_0 \oplus S \in \mathfrak{S}$ then $M \oplus S \in \mathcal{F}(n)$.

Conversely, suppose that $S_n \xrightarrow{\delta_n} \cdots \xrightarrow{\delta_1} S_0 \xrightarrow{\eta} M \oplus S \to 0$ is an n-stem where $M \oplus S \in \mathcal{F}(n)$. We may decompose this into a pair of exact sequences

$$S_n \xrightarrow{\delta_n} \cdots \xrightarrow{\delta_2} S_1 \xrightarrow{p} K \to 0 \qquad (*)$$

$$0 \to K \xrightarrow{i} S_0 \xrightarrow{\eta} M \oplus S \to 0 \qquad (**)$$

where $\delta_1 = i \circ p$. Take $\pi_1 : M \oplus S \to M$, $\pi_2 : M \oplus S \to S$ to be the canonical projections and put $S' = \mathrm{Ker}(\pi_2 \circ \epsilon)$. As in the proof of $\mathcal{P}(0)$, $S' \in \mathfrak{S}$ and $\pi_1 \circ \epsilon : S' \to M$ is epimorphic. Moreover there is an isomorphism of K' with $\mathrm{Ker}(\pi_1 \circ \eta : S' \to M)$ giving an exact sequence

$$0 \to K' \to S' \xrightarrow{\pi_1 \circ \eta} M \to 0. \qquad (***)$$

Splicing $(***)$ with $(*)$ gives an n-stem $S_n \xrightarrow{\delta_n} \cdots \xrightarrow{\delta_1} S' \xrightarrow{\pi_1 \circ \eta} M \to 0$; hence $M \in \mathcal{F}(n)$. This completes the proof. $\square$

Corollary 6.7. *If $M \sim M'$ then $M \in \mathcal{F}(n) \Leftrightarrow M' \in \mathcal{F}(n)$.*

In view of Corollary 6.7 we extend the condition $\mathcal{F}(n)$ from objects in $\mathfrak{A}$ to $\mathfrak{S}$-classes by saying that the $\mathfrak{S}$-class $[K]$ satisfies $\mathcal{F}(n)$ when K satisfies $\mathcal{F}(n)$.

7. A Strong Comparison Theorem for Syzygies

Given an n-stem $\mathbf{S}^{(n)} = (S_n \xrightarrow{\partial_n} S_{n-1} \xrightarrow{\partial_{n-1}} \cdots \xrightarrow{\partial_2} S_1 \xrightarrow{\partial_1} S_0 \xrightarrow{\epsilon} M \to 0)$ over M put $J_{r+1} = \operatorname{Ker}(\partial_r)$. Suppose given another n-stem over M

$$\widetilde{\mathbf{S}}^{(n)} = (\widetilde{S}_n \xrightarrow{\widetilde{\partial}_n} \widetilde{S}_{n-1} \xrightarrow{\widetilde{\partial}_{n-1}} \cdots \xrightarrow{\widetilde{\partial}_2} \widetilde{S}_1 \xrightarrow{\widetilde{\partial}_1} \widetilde{S}_0 \xrightarrow{\widetilde{\epsilon}} M \to 0)$$

with $\widetilde{J}_{r+1} = \operatorname{Ker}(\widetilde{\partial}_r)$ and suppose that $\varphi : \widetilde{\mathbf{S}}^{(n)} \to \mathbf{S}^{(n)}$ is a dominating homomorphism. From the results of Section 5, for $0 \leq r \leq n-2$ we obtain commutative diagrams $\mathcal{D}(r)$ in which all rows and columns are exact

$$\mathcal{D}(r) = \begin{cases} \begin{array}{ccccccccc} & & 0 & & 0 & & 0 & & \\ & & \downarrow & & \downarrow & & \downarrow & & \\ 0 \to & & V_{r+1} & \xrightarrow{\widehat{i_r}} & T_r & \xrightarrow{\widehat{p_r}} & V_r & \to 0 \\ & & \downarrow j_r^- & & \downarrow j_r & & \downarrow j_r^+ & & \\ 0 \to & & \widetilde{J}_{r+1} & \xrightarrow{\widetilde{i_r}} & \widetilde{S}_r & \xrightarrow{\widetilde{p_r}} & \widetilde{J}_r & \to 0 \\ & & \downarrow \varphi_r^- & & \downarrow \varphi_r & & \downarrow \varphi_r^+ & & \\ 0 \to & & J_{r+1} & \xrightarrow{i_r} & S_r & \xrightarrow{p_r} & J_r & \to 0 \\ & & \downarrow & & \downarrow & & \downarrow & & \\ & & 0 & & 0 & & 0 & & \end{array} \end{cases}$$

and in which $V_0 = 0$ and $J_0 = \widetilde{J}_0 = M$. As $0 \to T_r \to \widetilde{S}_r \to S_r \to 0$ is exact and $S_r, \widetilde{S}_r \in \mathfrak{S}$ it follows that $\widetilde{S}_r \cong T_r \oplus S_r$ and hence:

$$\text{Each } T_r \in \mathfrak{S}. \tag{7.1}$$

As $V_0 = 0$ then $V_1 = T_0$ so that:

$$V_1 \in \mathfrak{S}. \tag{7.2}$$

From the exact sequences $0 \to V_{r+1} \xrightarrow{\widehat{i_r}} T_r \xrightarrow{\widehat{p_r}} V_r \to 0$ it follows from $\mathfrak{S}(2)$ that

$$V_r \in \mathfrak{S} \quad \text{for } 1 \leq r \leq n-1. \tag{7.3}$$

Consequently,

$$0 \to V_{r+1} \xrightarrow{\widehat{i_r}} T_r \xrightarrow{\widehat{p_r}} V_r \to 0 \text{ splits for } 0 \leq r \leq n-2. \qquad (7.4)$$

Hence,

$$T_r \cong V_{r+1} \oplus V_r \quad \text{for } 0 \leq r \leq n-2. \qquad (7.5)$$

Theorem 7.1. *Let M be an object in $\mathcal{F}(n)$ and let $\mathbf{S}^{(n)}, \widetilde{\mathbf{S}}^{(n)}$ be n-stems over M. If $\mathbf{S}^{(n)} \preceq \widetilde{\mathbf{S}}^{(n)}$ then $\widetilde{J}_r$ splits over J_r for $1 \leq r \leq n-1$.*

Proof. By Theorem 5.2 we have diagrams $\mathcal{D}(r)$ with exact rows and columns for $0 \leq r \leq n-2$. First consider $\mathcal{D}(0)$

$$\mathcal{D}(0) = \begin{cases} & 0 \qquad\quad 0 \qquad\quad 0 \\ & \downarrow \qquad\quad \downarrow \qquad\quad \downarrow \\ 0 \to & V_1 \; = \; T_0 \xrightarrow{\widehat{p_0}} 0 \to 0 \\ & \downarrow j_0^- \quad\;\; \downarrow j_0 \qquad\;\; \downarrow \\ 0 \to & \widetilde{J}_1 \xrightarrow{\widetilde{i_0}} \widetilde{S}_0 \xrightarrow{\widetilde{\epsilon}} M \to 0 \\ & \downarrow \varphi_0^- \quad\;\; \downarrow \varphi_0 \qquad\;\; \| \text{ Id} \\ 0 \to & J_1 \xrightarrow{i_0} S_0 \xrightarrow{\epsilon} M \to 0 \\ & \downarrow \qquad\quad \downarrow \qquad\quad \downarrow \\ & 0 \qquad\quad 0 \qquad\quad 0 \end{cases}$$

By hypothesis we have that $S_0 \in \mathfrak{S}$ so that the middle column splits. If $\rho: \widetilde{S}_0 \to T_0$ is a left splitting of the middle column then $\rho \circ \widetilde{i}_0$ is a left splitting of the left-hand column and $\widetilde{J}_1 \cong J_1 \oplus V_1$. Suppose, inductively, that $0 \to V_r \xrightarrow{j_r^+} \widetilde{J}_r \xrightarrow{\varphi_r^+} J_r \to 0$ splits for $t \leq r \leq n-2$ and consider

$$\mathcal{D}(r) = \begin{cases} & 0 \qquad\qquad 0 \qquad\qquad 0 \\ & \downarrow \qquad\qquad \downarrow \qquad\qquad \downarrow \\ 0 \to & V_{r+1} \xrightarrow{\widehat{i_r}} T_r \xrightarrow{\widehat{p_r}} V_r \to 0 \\ & \downarrow j_r^- \qquad\; \downarrow j_r \qquad\; \downarrow j_r^+ \\ 0 \to & \widetilde{J}_{r+1} \xrightarrow{\widetilde{i_r}} \widetilde{S}_r \xrightarrow{\widetilde{p_r}} \widetilde{J}_r \to 0 \\ & \downarrow \varphi_r^- \qquad\; \downarrow \varphi_r \qquad\; \downarrow \varphi_r^+ \\ 0 \to & J_{r+1} \xrightarrow{i_r} S_r \xrightarrow{p_r} J_r \to 0 \\ & \downarrow \qquad\qquad \downarrow \qquad\qquad \downarrow \\ & 0 \qquad\qquad 0 \qquad\qquad 0 \end{cases}$$

As $r \leq n-2$ then we see from (7.3) that $V_r, V_{r+1} \in \mathfrak{S}$. Moreover $S_r \in \mathfrak{S}$ so that both S_r and V_r are projective. It follows from Theorem 4.1 that the sequence $0 \to V_{r+1} \to \widetilde{J}_{r+1} \to J_{r+1} \to 0$ splits and so $\widetilde{J}_{r+1} \cong J_{r+1} \oplus V_{r+1}$. As $V_{r+1} \in \mathfrak{S}$ this completes the proof. □

Corollary 7.2. *Let $\mathbf{S}^{(n)}$ and $\widetilde{\mathbf{S}}^{(n)}$ be n-stems over M with syzygies $(J_r)_{1 \leq r \leq n}$ $(\widetilde{J}_r)_{1 \leq r \leq n}$ respectively. If $\mathbf{S}^{(n)}$ is minimal then $\widetilde{J}_r$ splits over J_r for $1 \leq r \leq n-1$.*

Thus we have proved Theorem 1.3 of the Introduction.

In the limiting case, an ∞-stem will be called a *complete $\mathfrak{S}$-resolution* of M. The statement of Corollary 7.2 is then modified as follows.

Corollary 7.3. *Let $\mathbf{S}$ and $\widetilde{\mathbf{S}}$ be complete $\mathfrak{S}$-resolutions of M with syzygies $(J_r)_{1 \leq r}, (\widetilde{J}_r)_{1 \leq r}$. If $\mathbf{S}$ is minimal then $\widetilde{J}_r$ splits over J_r for all r.*

8. Uniqueness of Minimal Resolutions

Let $M \in \mathcal{F}(n)$ and let $\mathbf{S}^{(n)}$ be an n-stem over M

$$\mathbf{S}^{(n)} = (S_n \xrightarrow{\partial_n} S_{n-1} \xrightarrow{\partial_{n-1}} \cdots \xrightarrow{\partial_2} S_1 \xrightarrow{\partial_1} S_0 \xrightarrow{\epsilon} M \to 0).$$

We say that $\mathbf{S}^{(n)}$ is a *minimal n-stem* when, given any other n-stem $\widetilde{\mathbf{S}}$ over M, there is a dominating morphism $\varphi : \widetilde{\mathbf{S}}^{(n)} \to \mathbf{S}^{(n)}$ over Id_M thus

$$\begin{array}{c}\widetilde{\mathbf{S}}^{(n)} \\ \varphi \downarrow \\ \mathbf{S}^{(n)}\end{array} = \left(\begin{array}{ccccccc} \widetilde{S}_n & \xrightarrow{\widetilde{\partial}_n} & \cdots & \xrightarrow{\widetilde{\partial}_1} & \widetilde{S}_0 & \xrightarrow{\widetilde{\epsilon}} & M & \to 0 \\ \varphi_n \downarrow & & & & \varphi_0 \downarrow & & \downarrow \mathrm{Id}_M & \\ S_n & \xrightarrow{\partial_n} & \cdots & \xrightarrow{\partial_1} & S_0 & \xrightarrow{\epsilon} & M & \to 0 \end{array}\right).$$

In particular, φ_r is epimorphic for each r. A straightforward deduction from Corollaries 7.2 and 6.2 then shows the following.

Proposition 8.1. *If $\mathbf{S}^{(n)}, \widetilde{\mathbf{S}}^{(n)}$ are both minimal n-stems over M, then $\mathbf{S}^{(n)} \cong_{\mathrm{Id}_M} \widetilde{\mathbf{S}}^{(n)}$.*

This is easily strengthened to allow variation of the differentials as follows.

Proposition 8.2. *Suppose given n-stems over M as follows:*

$$\mathbf{S}^{(n)} = \left(S_n \xrightarrow{\partial_n} \cdots \xrightarrow{\partial_1} S_0 \xrightarrow{\epsilon} M \to 0\right);$$

$$\widehat{\mathbf{S}}^{(n)} = \left(S_n \xrightarrow{\delta_n} \cdots \xrightarrow{\delta_1} S_0 \xrightarrow{\eta} M \to 0\right);$$

If $\mathbf{S}^{(n)}$ is minimal then so also is $\widehat{\mathbf{S}}^{(n)}$.

Let $M, M' \in \mathcal{F}(n)$ and let $\mathbf{S}^{(n)}$, $\mathbf{T}^{(n)}$ be n-stems over M, M' respectively:

$$\mathbf{S}^{(n)} = (S_n \stackrel{\partial_n}{\to} S_{n-1} \stackrel{\partial_{n-1}}{\to} \cdots \stackrel{\partial_2}{\to} S_1 \stackrel{\partial_1}{\to} S_0 \stackrel{\epsilon}{\to} M \to 0);$$

$$\mathbf{T}^{(n)} = (T_n \stackrel{\partial'_n}{\to} T_{n-1} \stackrel{\partial'_{n-1}}{\to} \cdots \stackrel{\partial'_2}{\to} T_1 \stackrel{\partial'_1}{\to} T_0 \stackrel{\eta}{\to} M' \to 0).$$

We may form an n-stem $\mathbf{S}^{(n)} \oplus \mathbf{T}^{(n)}$ over $M \oplus M'$ thus

$$S_n \oplus T_n \stackrel{\delta_n}{\to} S_{n-1} \oplus T_{n-1} \stackrel{\delta_{n-1}}{\to} \cdots \stackrel{\delta_2}{\to} S_1 \oplus T_1 \stackrel{\delta_1}{\to} S_0 \oplus T_0 \stackrel{\epsilon \oplus \eta}{\to} M \oplus M' \to 0,$$

where $\delta_r = \begin{pmatrix} \partial_r & 0 \\ 0 & \partial'_r \end{pmatrix}$. An n-stem $\mathbf{T}^{(n)}$ over 0 is simply an exact sequence $\mathbf{T}^{(n)} = (T_n \to T_{n-1} \to \cdots \to T_1 \to T_0 \to 0)$ where each $T_r \in \mathfrak{S}$. Moreover, $\mathbf{S}^{(n)} \oplus \mathbf{T}^{(n)}$ is then an n-stem over $M \oplus 0 \cong M$. We now have the following theorem which is Theorem 1.2 of the Introduction.

Theorem 8.3. *If $\mathbf{S}^{(n)}, \widetilde{\mathbf{S}}^{(n)}$ are n-stems over M and $\mathbf{S}^{(n)}$ is minimal, then $\widetilde{\mathbf{S}}^{(n-1)} \cong_{\mathrm{Id}_M} \mathbf{S}^{(n-1)} \oplus \mathbf{T}^{(n-1)}$ for some $(n-1)$-stem $\mathbf{T}^{(n-1)}$ over 0.*

Proof. Given a dominating homomorphism $\varphi : \widetilde{\mathbf{S}}^{(n)} \to \mathbf{S}^{(n)}$ we construct, as in (5.2), a commutative diagram in which all rows and columns are exact

$$\begin{cases}
& & 0 & & 0 & & & & 0 & & 0 & \\
& & \downarrow & & \downarrow & & & & \downarrow & & \downarrow & \\
& & T_{n-1} & \stackrel{\widehat{\partial}_{n-1}}{\to} & T_{n-2} & \stackrel{\widehat{\partial}_{n-2}}{\to} \cdots \stackrel{\widehat{\partial}_1}{\to} & T_0 & \to & 0 & \to & 0 \\
& & j_{n-1} \downarrow & & j_{n-2} \downarrow & & & & j_0 \downarrow & & \downarrow & \\
\widetilde{S}_n & \stackrel{\widetilde{\partial}_n}{\to} & \widetilde{S}_{n-1} & \stackrel{\widetilde{\partial}_{n-1}}{\to} & \widetilde{S}_{n-2} & \stackrel{\widetilde{\partial}_{n-2}}{\to} \cdots \stackrel{\widetilde{\partial}_1}{\to} & \widetilde{S}_0 & \stackrel{\widetilde{\epsilon}}{\to} & M & \to & 0 \\
\varphi_n \downarrow & & \varphi_{n-1} \downarrow & & \varphi_{n-2} \downarrow & & & & \varphi_0 \downarrow & & \downarrow & \\
S_n & \stackrel{\partial_n}{\to} & S_{n-1} & \stackrel{\partial_{n-1}}{\to} & S_{n-2} & \stackrel{\partial_{n-2}}{\to} \cdots \stackrel{\partial_1}{\to} & S_0 & \stackrel{\epsilon}{\to} & M & \to & 0 \\
\downarrow & & \downarrow & & \downarrow & & & & \downarrow & & \downarrow & \\
0 & & 0 & & 0 & & & & 0 & & 0 &
\end{cases}$$

In particular we have an $(n-1)$-stem over the zero object, namely

$$\mathbf{T} = (T_{n-1} \stackrel{\widehat{\partial}_{n-1}}{\to} T_{n-2} \stackrel{\widehat{\partial}_{n-2}}{\to} \cdots \stackrel{\widehat{\partial}_1}{\to} T_0 \to 0 \to 0).$$

For $0 \leq k \leq n-2$ we obtain commutative diagrams $\mathcal{D}(k)$ as in Section 5 in which all rows and columns are exact. As the right-hand column of $\mathcal{D}(0)$ is trivially split and both $\widetilde{S}_0$ and 0 are projective we may, by Theorem 4.1

construct a complete splitting $(r_0^+, r_0, 0)$ of $\mathcal{D}(0)$ as follows:

$$\begin{array}{ccccc}
V_1 & = & T_0 & \to & 0 \\
\uparrow r_0^+ & & \uparrow r_0 & & \uparrow \\
J_1 & \xrightarrow{\tilde{i}_0} & \tilde{S}_0 & \xrightarrow{\tilde{\epsilon}} & M
\end{array}$$

Next consider $\mathcal{D}(1)$, recalling that the right-hand column of $\mathcal{D}(1)$ is identical with the left-hand column of $\mathcal{D}(0)$. Defining $r_1^- = r_0^+$ we see that r_1^- is a (left) splitting of the right-hand column of $\mathcal{D}(1)$. As V_1 and S_1 are projective then, by Theorem 4.1, r_1^- extends to a complete splitting (r_1^-, r_1, r_1^+) of $\mathcal{D}(1)$.

Suppose inductively that for $t \leq k-1$ we have constructed complete splittings (r_t^+, r_t, r_t^-) of $\mathcal{D}(t)$ in such a way that $r_t^- = r_{t-1}^+$. Defining $r_k^- = r_{k-1}^+$ gives a splitting of the right-hand column of $\mathcal{D}(k)$. Now $S_k \in \mathfrak{S}$ by hypothesis and we have seen in (7.3) that $V_k \in \mathfrak{S}$; thus both S_k and V_k are projective. It again follows from Theorem 4.1 that we may extend r_k^- to a complete splitting (r_k^+, r_k, r_k^-) of $\mathcal{D}(k)$.

Inductively, for k in the range $0 \leq k \leq n-2$, we construct complete splittings (r_k^+, r_k, r_k^-) of $\mathcal{D}(k)$ such that $r_k^- = r_{k-1}^+$ when $k \geq 1$. Finally, applying Theorem 4.1 to

$$\mathcal{E} = \begin{cases} 0 \to T_{n-1} \xrightarrow{j_1} \tilde{S}_{n-1} \xrightarrow{\varphi_1} S_{n-1} \to 0 \\ \quad\quad \hat{p}_{n-1} \downarrow \quad\quad \tilde{p}_{n-1} \downarrow \quad\quad p_{n-1} \downarrow \\ 0 \to V_{n-1} \xrightarrow{j_0} \tilde{J}_{n-1} \xrightarrow{\varphi_0} J_{n-1} \to 0 \end{cases}$$

we may construct a left splitting $r_{n-1} : \tilde{S}_{n-1} \to T_{n-1}$ of the exact sequence

$$0 \to T_{n-1} \xrightarrow{j_{n-1}^1} \tilde{S}_{n-1} \xrightarrow{\varphi_{n-1}^1} S_{n-1} \to 0$$

making the following diagram commute:

$$\begin{array}{ccc}
T_{n-1} & \xleftarrow{r_1} & \tilde{S}_{n-1} \\
\downarrow \hat{p}_{n-1} & & \downarrow \tilde{p}_{n-1} \\
V_{n-1} & \xleftarrow{r_0} & \tilde{J}_{n-1}
\end{array}$$

It follows that we have constructed a morphism of exact sequences

$$\begin{array}{c} \tilde{\mathbf{S}}^{(n-1)} \\ \mathbf{r} \downarrow \\ \mathbf{T} \end{array} = \begin{cases} \tilde{S}_{n-1} \xrightarrow{\tilde{\partial}_{n-1}} \tilde{S}_{n-2} \xrightarrow{\tilde{\partial}_{n-2}} \cdots \xrightarrow{\tilde{\partial}_1} \tilde{S}_0 \xrightarrow{\tilde{\epsilon}} M \to 0 \\ r_{n-1} \downarrow \quad\quad r_{n-2} \downarrow \quad\quad\quad\quad r_0 \downarrow \quad\quad \downarrow \\ T_{n-1} \xrightarrow{\partial_{n-1}} T_{n-2} \xrightarrow{\partial_{n-2}} \cdots \xrightarrow{\partial_1} T_0 \to 0 \to 0 \end{cases}$$

Then $\binom{\varphi}{\mathbf{r}} : \tilde{\mathbf{S}}^{(n-1)} \to \mathbf{S}^{(n-1)} \oplus \mathbf{T}^{(n-1)}$ is the required isomorphism. $\square$

In the case of complete $\mathfrak{S}$-resolutions we may continue the construction of the complete splittings (r_k^+, r_k, r_k^-) indefinitely to obtain Theorem 1.1 of the Introduction.

Theorem 8.4. *Let* $\mathbf{S}$ *and* $\widetilde{\mathbf{S}}$ *be complete* $\mathfrak{S}$-*resolutions of* M. *If* $\mathbf{S}$ *is minimal, then* $\widetilde{\mathbf{S}} \cong \mathbf{S} \oplus \mathbf{T}$ *for some complete* $\mathfrak{S}$-*resolution* $\mathbf{T}$ *of* 0.

9. The Structure of the Stable Syzygies $\Omega_n(M)$

If $M \in \mathcal{F}(0)$ then there is an exact sequence $0 \to J \to S \to M \to 0$ with $S \in \mathfrak{S}$. We write

$$\Omega_1(M) = [J].$$

$\Omega_1(M)$ is called the *first stable syzygy* of M relative to $\mathfrak{S}$. It is well defined as, by Proposition 3.1, the $\mathfrak{S}$-class $[J]$ depends only upon M. Moreover:

$$\text{Let } M, M' \in \mathcal{F}(0); \text{ if } M \sim M' \text{ then } \Omega_1(M) = \Omega_1(M'). \tag{9.1}$$

More generally, if $M, M' \in \mathcal{F}(n)$ then there are exact sequences

$$0 \to J \to S_n \to \cdots \to S_0 \to M \to 0,$$

$$0 \to J' \to S'_n \to \cdots \to S'_0 \to M' \to 0$$

with $S_i, S'_j \in \mathfrak{S}$; (9.1) now generalizes straightforwardly to give:

$$\text{If } M \sim M' \text{ then } J \sim J'. \tag{9.2}$$

If $(0 \to J \to S_n \to \cdots \to S_0 \to M \to 0)$ is an n-stem over M we write

$$\Omega_{n+1}(M) = [J].$$

Here $\Omega_{n+1}(M)$ is the $(n+1)^{th}$-*stable syzygy of* M *relative to* $\mathfrak{S}$; to uniformize notation we shall write the stable class $[M]$ of M as $[M] = \Omega_0(M)$. From (9.2) we now obtain:

$$\text{Let } M, M' \in \mathcal{F}(n); \text{ if } M \sim M' \text{ then } \Omega_{n+1}(M) = \Omega_{n+1}(M'). \tag{9.3}$$

One sees easily that:

$$\text{If } M \in \mathcal{F}(n) \text{ then } \Omega_r(M) \text{ satisfies } \mathcal{F}(n-r) \text{ for } 1 \leq r \leq n. \tag{9.4}$$

If M satisfies $\mathcal{F}(n)$ then although $\Omega_{n+1}(M)$ is defined, it need not, in general, satisfy $\mathcal{F}(0)$. In this context, for $M \in \mathcal{F}(n)$ we see that:

$$\Omega_{n+1}(M) \text{ satisfies } \mathcal{F}(0) \iff M \text{ satisfies } \mathcal{F}(n+1). \tag{9.5}$$

10. Realizing Elements of $\Omega_n(M)$ as Syzygies

We say that $M \in \mathfrak{A}$ is *1-coprojective* when, for any $S \in \mathfrak{S}$, any exact sequence of the form $0 \to S \to X \to M \to 0$ splits; then:

If $M \sim M'$ then M is 1-coprojective $\iff$ M' is 1-coprojective. (10.1)

We have the following 'realization lemma' (cf. [3, p. 107]):

If M is a 1-coprojective of type $\mathcal{F}(0)$ then any $J \in \Omega_1(M)$ occurs in an exact sequence $0 \to J \to S \to M \to 0$ in which $S \in \mathfrak{S}$. (10.2)

More generally, we say that $M \in \mathfrak{A}$ is $(n+1)$-*coprojective* when $\Omega_r(M)$ is defined and 1-coprojective for $0 \le r \le n$.

Proposition 10.1. *Suppose that $M \in \mathcal{F}(n)$ is $(n+1)$-coprojective; then for any sequence $(J_r)_{1 \le r \le n+1}$ with $J_r \in \Omega_r(M)$ there exists an n-stem*

$$\mathbf{S} = \left(S_n \xrightarrow{\partial_n} S_{n-1} \xrightarrow{\partial_{n-1}} \cdots \xrightarrow{\partial_1} S_0 \xrightarrow{\epsilon} M \to 0 \right)$$

in which $J_r \cong \mathrm{Ker}(\partial_{r-1})$ for $1 \le r \le n+1$.

Proof. By induction on n. Taking $J_1 = J$ and putting $\partial_0 = \epsilon$ then the statement for $n = 0$ is simply (10.2). Thus suppose that $n = 1$ and let $J_1 \in \Omega_1(M)$, $J_2 \in \Omega_2(M)$; by (10.2) there exist an object $S_0 \in \mathfrak{S}$ and an exact sequence $0 \to J_1 \xrightarrow{i_0} S_0 \xrightarrow{\epsilon} M \to 0$. The hypothesis $M \in \mathcal{F}(1)$ implies that $J_1 \in \mathcal{F}(0)$. As $\Omega_2(M) = \Omega_1(J_1)$ then $J_2 \in \Omega_1(J_1)$ so we may apply (10.2) to obtain an exact sequence $0 \to J_2 \xrightarrow{i_1} S_1 \xrightarrow{\pi_1} J_1 \to 0$ where $S_1 \in \mathfrak{S}$. Splicing these two sequences together by putting $\partial_1 = i_0 \circ \pi_1$ we obtain an exact sequence $(0 \to J_2 \xrightarrow{i_1} S_1 \xrightarrow{\partial_1} S_0 \xrightarrow{\epsilon} M \to 0)$ where $S_0, S_1 \in \mathfrak{S}$. Then $(S_1 \xrightarrow{\partial_1} S_0 \xrightarrow{\epsilon} M \to 0)$ is a 1-stem with the stated properties.

In general, suppose proved for $n-1$ where $n \ge 2$ and let $(J_r)_{1 \le r \le n+1}$ be a sequence with $J_r \in \Omega_r(M)$. By hypothesis there exists an $(n-1)$-stem

$$\mathbf{S}' = \left(S_{n-1} \xrightarrow{\partial_{n-1}} \cdots \xrightarrow{\partial_1} S_0 \xrightarrow{\epsilon} M \to 0 \right).$$

in which $S_0 \ldots S_{n-1} \in \mathfrak{S}$ and $J_r \cong \mathrm{Ker}(\partial_{r-1})$ for $1 \le r \le n$. We may write this in co-augmented form as

$$\mathbf{S}' = \left(0 \to J_n \xrightarrow{i_{n-1}} S_{n-1} \xrightarrow{\partial_{n-1}} \cdots \xrightarrow{\partial_1} S_0 \xrightarrow{\epsilon} M \to 0 \right),$$

The hypothesis $M \in \mathcal{F}(n)$ implies that $J_n \in \mathcal{F}(0)$. As $\Omega_{n+1}(M) = \Omega_1(J_n)$ we see that $J_{n+1} \in \Omega_1(J_n)$. Apply (10.2) again to obtain an exact sequence

$$0 \to J_{n+1} \overset{i_n}{\to} S_n \overset{\pi_n}{\to} J_n \to 0,$$

where $S_n \in \mathfrak{S}$. Splicing these last two sequences together gives an n-stem with the stated properties

$$0 \to J_{n+1} \overset{i_n}{\to} S_n \overset{\partial_n}{\to} \cdots \overset{\partial_1}{\to} S_0 \overset{\epsilon}{\to} M \to 0, \qquad (*)$$

where $\partial_n = i_{n-1} \circ \pi_n$. □

We say that $\Omega_r(M)$ is *relatively straight* when there exists $N_0 \in \Omega_r(M)$ such that any other $N \in \Omega_r(M)$ may be written in the form $N \cong N_0 \oplus T$ for some $T \in \mathfrak{S}$. We note the following consequence of minimality.

Theorem 10.2. *Suppose that $M \in \mathcal{F}(n+1)$ admits a minimal $(n+1)$-stem $\mathbf{S}^{(n+1)}$. If $\Omega_{n-1}(M)$ is 1-coprojective then $\Omega_n(M)$ is relatively straight.*

Proof. Write $\mathbf{S}^{(n+1)} = (S_{n+1} \overset{\partial_{n+1}}{\to} S_n \overset{\partial_n}{\to} \cdots \overset{\partial_1}{\to} S_0 \overset{\epsilon}{\to} M \to 0)$ and for $1 \leq r \leq n+1$ put $J_r = \mathrm{Im}(\partial_r)$. Choose $J \in \Omega_n(M)$. We must show that there exists $T \in \mathfrak{S}$ such that $J \cong J_n \oplus T$. Write the truncation $\mathbf{S}^{(n-2)}$ in co-augmented form

$$\mathbf{S}^{(n-2)} = \left(J_{n-1} \overset{i_{n-2}}{\to} S_{n-2} \overset{\partial_{n-2}}{\to} \cdots \overset{\partial_1}{\to} S_0 \overset{\epsilon}{\to} M \to 0 \right).$$

Clearly $J \in \Omega_1(J_{n-1})$ as $\Omega_n(M) = \Omega_1(J_{n-1})$ and, by hypothesis, J_{n-1} is 1-coprojective. Thus by (10.1), there exists an exact sequence

$$\mathbf{E} = (0 \to J \to E_0 \to J_{n-1} \to 0),$$

where $E_0 \in \mathfrak{S}$. As $M \in \mathcal{F}(n+1)$ and $J \in \Omega_n(M)$ then $J \in \mathcal{F}(1)$ so there exists a 1-stem $\mathbf{F} = (F_1 \to F_0 \to J \to 0)$. Yoneda product $\mathbf{F} \circ \mathbf{E} \circ \mathbf{S}^{(n-2)}$ gives an $(n+1)$-stem

$$\widetilde{\mathbf{S}}^{(n+1)} = \left(F_1 \overset{\delta_{n+1}}{\to} F_0 \overset{\delta_n}{\to} E_0 \overset{\delta_{n-1}}{\to} \cdots \overset{\delta_1}{\to} \widetilde{S}_0 \overset{\epsilon}{\to} M \to 0 \right)$$

where $\widetilde{S}_r = S_r$ for $r \leq n-2$ and where $J = \mathrm{Im}(\delta_n)$. As $\mathbf{S}^{(n+1)}$ is minimal it follows from Corollary 7.2 that J splits over $J_n = \mathrm{Im}(\partial_n)$. Thus, as claimed, there exists $T \in \mathfrak{S}$ such that $J \cong J_n \oplus T$. □

11. Minimal Epimorphisms

We define a category $\mathfrak{S}_{(-)}$ in which the objects are pairs (S, ϵ) where $S \in \mathfrak{S}$ and where ϵ is an epimorphism in $\mathfrak{A}$ with domain S and whose codomain is some, as yet unspecified, object in $\mathfrak{A}$. Morphisms in $\mathfrak{S}_{(-)}$ are then commutative squares of morphisms in $\mathfrak{A}$ thus:

$$\begin{array}{ccc} S' & \xrightarrow{\epsilon'} & M' \\ \varphi \downarrow & & \downarrow \varphi_- \\ S & \xrightarrow{\epsilon} & M. \end{array}$$

In this case we say that φ is a *morphism over* φ_-. In practice we shall only consider the case where φ_- is an isomorphism and usually, though not always, we shall take φ_- to be the identity morphism. For $M \in \mathfrak{A}$ we define a subcategory $\mathfrak{S}_M$ of $\mathfrak{S}_{(-)}$ by restricting morphisms to be commutative squares of the form

$$\begin{array}{ccc} S' & \xrightarrow{\epsilon'} & M \\ \varphi \downarrow & & \downarrow \mathrm{Id}_M \\ S & \xrightarrow{\epsilon} & M. \end{array}$$

If (S, ϵ), (S', ϵ') are objects in $\mathfrak{S}_M$ we write $(S, \epsilon) \preceq (S', \epsilon')$ whenever there exists a morphism $\varphi : (S', \epsilon') \to (S, \epsilon)$ in which $\varphi : S' \to S$ is an epimorphism in $\mathfrak{A}$. It is straightforward to observe that:

If $(S, \epsilon) \preceq (S', \epsilon')$ and $(S', \epsilon') \preceq (S'', \epsilon'')$ then $(S, \epsilon) \preceq (S'', \epsilon'')$. (11.1)

Slightly more subtle is

$$(S, \epsilon) \preceq (S', \epsilon') \wedge (S', \epsilon') \preceq (S, \epsilon) \iff (S, \epsilon) \cong (S', \epsilon'). \tag{11.2}$$

It follows that:

The relation '$\preceq$' induces a partial ordering on the isomorphism classes of $\mathfrak{S}_M$. (11.3)

For $E \in \mathfrak{S}$ we define the *base stabilisation* functor $\beta_E : \mathfrak{S}_M \to \mathfrak{S}_{M \oplus E}$ thus:

$$\beta_E \begin{pmatrix} S' & \xrightarrow{\epsilon'} & M \\ \varphi \downarrow & & \downarrow \mathrm{Id}_M \\ S & \xrightarrow{\epsilon} & M \end{pmatrix} = \begin{pmatrix} S' \oplus E & \xrightarrow{\epsilon' \oplus \mathrm{Id}} & M \oplus E \\ \varphi \oplus \mathrm{Id}_E \downarrow & & \downarrow \mathrm{Id}_M \\ S \oplus E & \xrightarrow{\epsilon \oplus \mathrm{Id}} & M \oplus E \end{pmatrix};$$

that is, β_E acts on objects by $\beta_E(S, \epsilon) = (S \oplus E, \epsilon \oplus \mathrm{Id}_M)$ and on morphisms by $\beta_E(\varphi) = \varphi \oplus \mathrm{Id}_E$. Observe that β_E is order preserving:

$$(S, \epsilon) \preceq (S', \epsilon') \implies \beta_E(S, \epsilon) \preceq \beta_E(S', \epsilon'). \tag{11.4}$$

Write $(S, \epsilon) \in \mathfrak{S}_{M \oplus E}$ as an exact sequence $0 \to K \hookrightarrow S \xrightarrow{\epsilon} M \oplus E \to 0$ and put $T = \text{Ker}(\pi_E \circ \epsilon)$ where $\pi_E : M \oplus E \to E$ is the canonical projection. We obtain a pair of exact sequences

$$0 \to T \to S \to S/T \to 0; \quad 0 \to T/K \to S/K \to S/T \to 0.$$

and a Noether isomorphism $S/T \cong (M \oplus E)/M \cong E$. In particular, $S/T \in \mathfrak{S}$. We may assemble the above into a commutative diagram with exact rows and columns

$$\left\{ \begin{array}{ccccccccc}
 & & 0 & & 0 & & & & \\
 & & \downarrow & & \downarrow & & & & \\
0 \to & & K & = & K & \to & 0 & & \\
 & & \downarrow & & \downarrow & & \downarrow & & \\
0 \longrightarrow & & T & \to & S & \xrightarrow{\tilde{\pi}} & S/T & \longrightarrow & 0 \\
 & & \downarrow \nu' & & \downarrow \nu & & \| \text{ Id} & & \\
0 \longrightarrow & & T/K & \to & S/K & \xrightarrow{\pi} & S/T & \longrightarrow & 0 \\
 & & \downarrow & & \downarrow & & \downarrow & & \\
 & & 0 & & 0 & & 0 & &
\end{array} \right. \qquad (*)$$

in which $\nu, \nu', \tilde{\pi}$ and π are all canonical morphisms. As S and $S/T \cong E$ are both in $\mathfrak{S}$, it follows from the middle row of (*) and property $\mathfrak{S}(2)$ that $T \in \mathfrak{S}$. As $S/T \cong E$ is projective we may choose a morphism $\tilde{\sigma} : S/T \to S$ which splits the middle row of (*) on the right. Now define $\sigma = \nu \circ \tilde{\sigma} : S/T \to S/K$. As $\pi \circ \nu = \tilde{\pi}$ we see that

$$\pi \circ \sigma = \text{Id}_{S/T}. \qquad (**)$$

That is, σ splits the bottom row of (*) on the right. Taking the corresponding left splittings $\tilde{\lambda} = \text{Id}_S - \tilde{\sigma}\tilde{\pi}$; $\lambda = \text{Id}_S - \sigma\pi$, one verifies easily that $\lambda \circ \nu = \nu' \circ \tilde{\lambda}$. In addition to (*) we have another commutative diagram with exact rows and columns

$$\left\{ \begin{array}{ccccccccc}
 & & 0 & & 0 & & & & \\
 & & \downarrow & & \downarrow & & & & \\
0 \leftarrow & & K & = & K & \leftarrow & 0 & & \\
 & & \downarrow & & \downarrow & & \downarrow & & \\
0 \longleftarrow & & T & \xleftarrow{\tilde{\lambda}} & S & \xleftarrow{\tilde{\sigma}} & S/T & \longleftarrow & 0 \\
 & & \downarrow \nu' & & \downarrow \nu & & \| \text{ Id} & & \\
0 \longleftarrow & & T/K & \xleftarrow{\lambda} & S/K & \xleftarrow{\sigma} & S/T & \longleftarrow & 0 \\
 & & \downarrow & & \downarrow & & \downarrow & & \\
 & & 0 & & 0 & & 0 & &
\end{array} \right. \qquad (***)$$

Thus there exists a Noether isomorphism $\natural_1 : (S, \nu) \xrightarrow{\simeq} (S, \epsilon)$ for some $(S, \nu) \in \mathfrak{S}_{S/K}$. As T and S/T are both in $\mathfrak{S}$, then $(T, \nu') \in \mathfrak{S}_{T/K}$ and $\beta_{S/T}(T, \nu')$ is well defined. Now consider the isomorphisms

$$\widetilde{h} : S \to T \oplus S/T; \quad h : S/K \to T/K \oplus S/T,$$
$$\widetilde{h}(x) = (\widetilde{\lambda}(x), \widetilde{\pi}(x)); \quad h(x) = (\lambda(x), \pi(x)).$$

Then $\widetilde{h}$ defines an isomorphism $\widetilde{h} : (S, \nu) \xrightarrow{\simeq}_h \beta_{S/T}(T, \nu')$ over h and there is another Noether isomorphism $\natural_2 : \beta_{S/T}(T, \nu') \xrightarrow{\simeq} \beta_E(T, \eta)$ where $\eta = \epsilon_{|T} : T \to M$. The composition $\natural_2 \circ \widetilde{h} \circ \natural_1^{-1} : (S, \epsilon) \xrightarrow{\simeq} \beta_E(T, \eta)$ is an isomorphism over $\mathrm{Id}_{M \oplus E}$. We have shown the following.

Theorem 11.1. $\beta_E : \mathfrak{S}_M \to \mathfrak{S}_{M \oplus E}$ *is surjective on isomorphism classes.*

For (S, ϵ), (S', ϵ') in $\mathfrak{S}_M$ consider morphisms $\varphi : \beta_E(S', \epsilon') \to \beta_E(S, \epsilon)$ in $\mathfrak{S}_{M \oplus E}$. Any such morphism is, at least, a morphism $\varphi : S' \oplus E \to S \oplus E$ in $\mathfrak{A}$ and so may be described as a matrix of $\mathfrak{A}$-morphisms

$$\varphi = \begin{pmatrix} A & B \\ C & D \end{pmatrix} \quad \text{where} \quad \begin{cases} A : S' \to S; & B : E \to S; \\ C : S' \to E; & D : E \to E. \end{cases}$$

A straightforward calculation shows that there is a one-to-one correspondence

$$\mathrm{Hom}_{\mathfrak{S}_{(-)}}(\beta_E(S', \epsilon'), \beta_E(S, \epsilon))$$
$$\xleftrightarrow{\simeq} \left\{ \begin{pmatrix} A & B \\ 0 & \mathrm{Id}_E \end{pmatrix} \; \middle| \; \begin{matrix} A \in \mathrm{Hom}_{\mathfrak{S}_{(-)}}((S', \epsilon'), (S, \epsilon)) \\ B \in \mathrm{Hom}_{\mathfrak{A}}(E, \mathrm{Ker}(\epsilon)) \end{matrix} \right\}.$$

Suppose given an $\mathfrak{A}$-morphism $\varphi = \begin{pmatrix} A & B \\ 0 & \mathrm{Id}_E \end{pmatrix} : S' \oplus E \to S \oplus E$. If A is epimorphic then so is φ. Conversely if φ is epimorphic, then since $S \oplus E$ is projective there exists an $\mathfrak{A}$-morphism $\sigma : S \oplus E \to S' \oplus E$ such that $\varphi \circ \sigma = \mathrm{Id}_{S \oplus E}$. Writing $\sigma = \begin{pmatrix} \sigma_{11} & \sigma_{12} \\ \sigma_{21} & \sigma_{22} \end{pmatrix}$ it follows easily that $\sigma_{21} = 0$ and $A \circ \sigma_{11} = \mathrm{Id}_S$; hence A is also epimorphic. From this it follows that

$$\beta_E(S, \epsilon) \preceq \beta_E(S', \epsilon') \Longrightarrow (S, \epsilon) \preceq (S', \epsilon').$$

As a consequence we have the following corollary.

Corollary 11.2. *For any $E \in \mathfrak{S}$, β_E induces an order preserving bijection on isomorphism types $\beta_E : \mathfrak{S}_M \xrightarrow{\simeq} \mathfrak{S}_{M \oplus E}$.*

An epimorphism (S, ϵ) in $\mathfrak{S}_M$ is said to be *absolutely minimal* when $(S, \epsilon) \preceq (S', \epsilon')$ for each $(S', \epsilon') \in \mathfrak{S}_M$. We may verify directly that:

If $(S, \epsilon), (S', \epsilon')$ are both absolutely minimal over M then

$$(S, \epsilon) \cong (S', \epsilon'). \tag{11.5}$$

We say that $\mathcal{A}\mathrm{bs}(M)$ holds precisely when there exists an absolutely minimal epimorphism (S, ϵ) in $\mathfrak{S}_M$. By Corollary 11.2 satisfaction of this condition depends only upon the $\mathfrak{S}$-class of M; that is:

Corollary 11.3. *If $M \sim M'$ then $\mathcal{A}\mathrm{bs}(M)$ holds $\iff \mathcal{A}\mathrm{bs}(M')$ holds.*

This condition may be reformulated to say:

$$\mathcal{A}\mathrm{bs}(M) \text{ holds} \iff M \text{ admits a minimal 0-stem.} \tag{11.6}$$

12. An Existence Criterion

When $M \in \mathfrak{A}$ we define a subcategory $\mathfrak{S}(n)_M$ of $\mathfrak{S}(n)$ by restricting morphisms to be commutative diagrams of the form

$$\begin{array}{c} \widetilde{\mathbf{S}} \\ \varphi \downarrow \\ \mathbf{S} \end{array} = \begin{pmatrix} \widetilde{S}_n & \stackrel{\widetilde{\partial_n}}{\to} & \cdots & \stackrel{\widetilde{\partial_1}}{\to} & \widetilde{S}_0 & \stackrel{\widetilde{\eta}}{\to} & M & \to 0 \\ \varphi_n \downarrow & & & & \varphi_0 \downarrow & & \downarrow \mathrm{Id}_M & \\ S_n & \stackrel{\partial_n}{\to} & \cdots & \stackrel{\partial_1}{\to} & S_0 & \stackrel{\epsilon}{\to} & M & \to 0 \end{pmatrix}.$$

For $\mathbf{S}, \mathbf{S}', \mathbf{S}''$ in $\mathfrak{S}(n)_M$ we may generalize the results of Section 11 as follows:

If $\mathbf{S} \preceq \mathbf{S}'$ and $\mathbf{S}' \preceq \mathbf{S}''$ then $\mathbf{S} \preceq \mathbf{S}''$. $\tag{12.1}$

If $\mathbf{S} \preceq \mathbf{S}'$ and $\mathbf{S}' \preceq \mathbf{S}$ then $\mathbf{S} \cong \mathbf{S}'$. $\tag{12.2}$

The relation '$\preceq$' induces a partial ordering on the isomorphism classes of $\mathfrak{S}_M(n)$. $\tag{12.3}$

For $E \in \mathfrak{S}$ there is a base stabilisation functor $\beta_E : \mathfrak{S}(n)_M \to \mathfrak{S}(n)_{M \oplus E}$ which transforms

$$\begin{array}{c} \widetilde{\mathbf{S}} \\ \varphi \downarrow \\ \mathbf{S} \end{array} = \begin{pmatrix} \widetilde{S}_n & \stackrel{\widetilde{\partial_n}}{\to} & \cdots & \stackrel{\widetilde{\partial_1}}{\to} & \widetilde{S}_0 & \stackrel{\widetilde{\eta}}{\to} & M & \to 0 \\ \varphi_n \downarrow & & & & \varphi_0 \downarrow & & \downarrow \mathrm{Id}_M & \\ S_n & \stackrel{\partial_n}{\to} & \cdots & \stackrel{\partial_1}{\to} & S_0 & \stackrel{\epsilon}{\to} & M & \to 0 \end{pmatrix}$$

to

$$\begin{array}{c} \beta_E(\widetilde{\mathbf{S}}) \\ \beta_E(\varphi) \downarrow \\ \beta(\mathbf{S}) \end{array} = \begin{pmatrix} \widetilde{S}_n & \stackrel{\widetilde{\partial_n}}{\to} & \cdots & \stackrel{\widetilde{i \circ \partial_1}}{\to} & \widetilde{S}_0 \oplus E & \stackrel{\widetilde{\eta} \oplus \mathrm{Id}}{\to} & M \oplus E & \to 0 \\ \varphi_n \downarrow & & & & \varphi_0 \downarrow & & \downarrow \mathrm{Id}_M & \\ S_n & \stackrel{\partial_n}{\to} & \cdots & \stackrel{i \circ \partial_1}{\to} & S_0 \oplus E & \stackrel{\epsilon \oplus \mathrm{Id}}{\to} & M \oplus E & \to 0 \end{pmatrix},$$

where $\widetilde{i} : \widetilde{S}_0 \to \widetilde{S}_0 \oplus E$ and $i : S_0 \to S_0 \oplus E$ are the canonical morphisms.

β_E is order preserving; that is, $\mathbf{S} \preceq \mathbf{S}' \implies \beta_E(\mathbf{S}) \preceq \beta_E(\mathbf{S}')$. (12.4)

$\beta_E : \mathfrak{S}(n)_M \to \mathfrak{S}(n)_{M \oplus E}$ is surjective on isomorphism classes. (12.5)

If $\mathbf{S}$, $\widetilde{\mathbf{S}}$ are objects in $\mathfrak{S}(n)_M$ then $\beta_E(\mathbf{S}) \preceq \beta_E(\widetilde{\mathbf{S}}) \implies \mathbf{S} \preceq \widetilde{\mathbf{S}}$. (12.6)

β_E induces an order preserving bijection on isomorphism types

$$\beta_E : \mathfrak{S}(n)_M \xrightarrow{\simeq} \mathfrak{S}(n)_{M \oplus E}. \quad (12.7)$$

We have the following useful consequence of (12.7).

If $\mathbf{S}$ is a minimal n-stem over M, then $\beta_E(\mathbf{S})$ is a minimal n-stem over $M \oplus E$. (12.8)

We say that $\mathcal{M}\text{in}_n(M)$ *holds* when M admits a minimal n-stem. Note that the condition $\mathcal{M}\text{in}_0(M)$ is simply a re-statement of $\mathcal{A}\text{bs}(M)$. Moreover from (12.7) it follows immediately that:

If $M \sim M'$ then $\mathcal{M}\text{in}_n(M)$ holds $\iff \mathcal{M}\text{in}_n(M')$ holds. (12.9)

Thus satisfaction of the condition $\mathcal{M}\text{in}_n(M)$ depends only upon the $\mathfrak{S}$-class $[M]$ of $M \in \mathfrak{A}$. Observe that for $M \in \mathcal{F}(n)$ we have the following theorem.

Theorem 12.1. $\mathcal{A}\text{bs}(M) \wedge \mathcal{M}\text{in}_{n-1}(\Omega_1(M)) \implies \mathcal{M}\text{in}_n(M)$.

Proof. Let $\mathbf{S}^{(0)} = (0 \to K \xrightarrow{i} S_0 \xrightarrow{\epsilon} M \to 0)$ be a minimal 0-stem over M and let $\mathbf{S}' = (S'_{n-1} \xrightarrow{\delta_{n-1}} \cdots \xrightarrow{\delta_1} S'_0 \xrightarrow{\eta} K \to 0)$ be a minimal $(n-1)$-stem over K. After re-indexing thus $S_{r+1} = S'_r; \partial_{r+1} = \delta_r$ we may splice $\mathbf{S}'$ with $\mathbf{S}^{(0)}$ to obtain an n-stem

$$\mathbf{S} = \mathbf{S}' \circ \mathbf{S}^{(0)} = \left(S_n \xrightarrow{\partial_n} \cdots \xrightarrow{\partial_2} S_1 \xrightarrow{\partial_1} S_0 \xrightarrow{\epsilon} M \to 0 \right),$$

where $\partial_1 = i \circ \eta$. We claim that $\mathbf{S}$ is minimal; that is, given an n-stem $\widetilde{\mathbf{S}}$ over M we must produce a dominating morphism $\Psi : \widetilde{\mathbf{S}} \to \mathbf{S}$ over Id_M. Thus write $\widetilde{\mathbf{S}}$ as a Yoneda product $\widetilde{\mathbf{S}} = \widetilde{\mathbf{S}}' \circ \widetilde{\mathbf{S}}^{(0)}$ where

$$\widetilde{\mathbf{S}}' = \left(\widetilde{S}_n \xrightarrow{\widetilde{\partial}_n} \cdots \xrightarrow{\widetilde{\partial}_2} \widetilde{S}_1 \xrightarrow{\widetilde{\eta}} \widetilde{K} \to 0 \right); \quad \widetilde{\mathbf{S}}^{(0)} = \left(0 \to \widetilde{K} \xrightarrow{\widetilde{i}} \widetilde{S}_0 \xrightarrow{\widetilde{\epsilon}} M \to 0 \right).$$

Then there is a dominating morphism of 0-stems $\psi_0 : \widetilde{\mathbf{S}}^{(0)} \to \mathbf{S}^{(0)}$

$$(*) = \begin{cases} & 0   0 \\ & \downarrow   \downarrow \\ 0 \to & E \cong E \to 0 \\ & \downarrow j_- \downarrow j_0 \downarrow \\ 0 \to & \widetilde{K} \xrightarrow{\widetilde{i}} \widetilde{S}_0 \xrightarrow{\widetilde{\epsilon}} M \to 0 \\ & \downarrow \psi_- \phantom{\xrightarrow{i}} \downarrow \psi_0 \parallel \mathrm{Id} \\ 0 \to & K \xrightarrow{i} S_0 \xrightarrow{\epsilon} M \to 0 \\ & \downarrow   \downarrow \\ & 0   0 \end{cases}$$

Observe that $E \in \mathfrak{S}$ and $\widetilde{K} \cong K \oplus E$. Thus by (12.8) $\beta_E(\mathbf{S}')$ is a minimal $(n-1)$-stem over $\widetilde{K} \cong K \oplus E$. Hence there exists a dominating morphism $\psi' : \widetilde{\mathbf{S}}' \to \beta_E(\mathbf{S}')$. Composition with the canonical morphism $\pi : \beta_E(\mathbf{S}') \to \mathbf{S}'$ then takes the form

$$\pi \circ \psi' = \begin{pmatrix} \widetilde{S}_n & \xrightarrow{\widetilde{\partial}_n} & \cdots & \widetilde{S}_2 & \xrightarrow{\widetilde{\partial}_2} & \widetilde{S}_1 & \xrightarrow{\widetilde{p}_1} & K \oplus E & \to 0 \\ \psi'_n \downarrow & & & \psi'_2 \downarrow & & \pi \circ \psi'_1 \downarrow & & \pi \downarrow & \\ S_n & \xrightarrow{\partial_n} & \cdots & S_2 & \xrightarrow{\partial_2} & S_1 & \xrightarrow{p_1} & K & \to 0 \end{pmatrix}.$$

Rewriting $\widetilde{K} \cong K \oplus E$ we may splice $\pi \circ \psi'$ with $\psi^{(0)}$ to obtain a morphism over Id_M

$$\Psi = \begin{pmatrix} \widetilde{S}_n & \xrightarrow{\widetilde{\partial}_n} & \cdots & \widetilde{S}_1 & \xrightarrow{\widetilde{\partial}_1} & \widetilde{S}_0 & \xrightarrow{\widetilde{\epsilon}} & M & \to 0 \\ \Psi_n \downarrow & & & \Psi_1 \downarrow & & \Psi_0 \downarrow & & \mathrm{Id} \downarrow & \\ S_n & \xrightarrow{\partial_n} & \cdots & S_1 & \xrightarrow{\partial_1} & S_0 & \xrightarrow{\epsilon} & M & \to 0 \end{pmatrix},$$

where $\Psi_r = \pi \circ \psi'_r$ for $r = 1$ and $\Psi_r = \psi'_r$ otherwise. Thus each Ψ_r is epimorphic and Ψ is a dominating morphism. $\square$

From Theorem 12.1 we deduce our criterion for the existence of a minimal n-stem.

Theorem 12.2. *Let $M \in \mathcal{F}(n)$ and suppose that $\mathcal{A}\mathrm{bs}(\Omega_r(M))$ holds for $0 \leq r \leq n$; Then M admits a minimal n-stem.*

In conclusion, we point out that Eilenberg's results from [1] can all be accommodated under the aegis of Theorem 12.2. For example, when Λ is a local ring, we take $\mathfrak{A}$ to be the category of finitely generated Λ-modules and $\mathfrak{S} \subset \mathfrak{A}$ to be the subclass of finitely generated free modules. Likewise, when

Λ is semisimple, we take $\mathfrak{A}$ to be the category of locally finitely generated graded Λ-modules and $\mathfrak{S} \subset \mathfrak{A}$ to be the subclass of quasi-free modules. In either case, every such module M belongs to $\mathcal{F}(\infty)$ and satisfies $\mathrm{Abs}(M)$. Hence every such module has a complete minimal resolution. However, as we shall show elsewhere, there are many more examples of minimal resolutions which are excluded *a priori* from Eilenberg's framework.

References

[1] S. Eilenberg, Homological dimension and syzygies. *Ann. of Math.* **64**, 328–336 (1956).
[2] D. Hilbert, Über die theorie der algebraischen formen. *Math. Ann.* **36**, 473–539 (1890).
[3] F. E. A. Johnson, *Syzygies and Homotopy Theory*. Springer-Verlag (2011).
[4] S. Lubkin, Imbedding of abelian categories. *Trans. Amer. Math. Soc.* **97**, 410–417 (1960).
[5] B. Mitchell, *Theory of Categories*. Academic Press (1965).